Technological change and agrarian structure

A study of Bangladesh

Iftikhar Ahmed
Foreword by William Cline

International Labour Office   Geneva

ISBN 92-2-102543-8

First published 1981

Printed in Switzerland

FOREWORD

Despite generally rapid economic growth in the 1950s and 1960s, the developing countries frequently showed poor results in two areas: agricultural production, and equity in income distribution. After the adverse change in the international economy beginning in the 1970s and associated with the two oil price shocks of 1973-74 and 1979-80 and the entrenchment of stagflation in industrial countries, it has become all the more important for developing countries to make greater progress in these two areas. For many, greater agricultural production remains a major avenue towards efficient import substitution, and therefore a promising strategy for coping with high oil import costs and stagnant industrial country markets for their exports. Agricultural progress seems all the more urgent in view of the new lows to which world grain reserves were falling at the beginning of the 1970s. At the same time, the dreary prospects for economic growth, regularly lowered in successive assessments by international agencies in recent years, make it all the more essential to consider means of making income distribution more equitable. The policy appeal of trickle-down theory, let alone its political viability, is weaker when growth prospects are more bleak.

For these reasons the issues of agrarian structure, productivity and technological change are crucial for current development strategies. There is increasing evidence that, in many developing countries, placing greater emphasis on a small-farm agrarian development strategy would improve both equity and agricultural productivity. Systematic studies have shown that small farms tend to achieve higher production relative to their available land, and employ relatively more labour and less capital, than large farms. As a result, the smaller farms tend to be more efficient in social terms than the large farms, considering the low social opportunity cost of labour and the high social cost of land and capital in most developing countries. The economic forces causing this divergence include the intensive use of family labour on small farms in a surplus labour context in which outside wage employment is scarce, and the fact that large landholders often impute a low price to the use of land resources (in part because they hold land as a portfolio asset hedge against inflation). These tendencies are reinforced by imperfect capital markets that tend to make the price of capital equipment lower for large farms than for small.

Compared with agrarian structures based on large estates, therefore, small-farm structures hold out the promise of greater productivity. Moreover, because they are typically labour intensive, they can lead to more equitable income distribution. At one extreme, the policy options for pursuing small-farm strategies include sweeping land reform. It is generally accepted that the thorough land reform in the Republic of Korea was a crucial factor in the country's achievement of a relatively equitable income distribution. More modest policies include a greater focus on official credit and other programmes for the existing small-farm sector, combined with land taxes designed to encourage the break-up of large estates with relatively underused land.

This policy orientation has been advocated by international agencies and supported by empirical documentation from scholarly studies. Yet the national authorities in agricultural and economic ministries frequently prefer to leave existing small farms at the margin of development schemes, and to shudder at the very thought of meaningful land reform. Their preference for a large-farm development strategy is partly the result of interest-group politics; the landed class remains powerful in many countries. However, their reluctance to adopt small-farm strategies stems partly from a technocratic belief that the large, mechanised modern farm points the way to the future.

Empirical studies such as Iftikhar Ahmed's analysis of the agrarian structure in Bangladesh can provide important evidence to alert policy makers about the developmental potential of small-farm strategies. Ahmed has built upon a growing literature on the relationship of farm size to productivity and technological change, using a detailed data set that enables him to explore these issues more fully than has usually been possible in previous studies. His results reinforce the conclusions of earlier studies that output per unit of available land, and labour intensity in production, are inversely related to farm size, even when compensating for factors such as land quality and availability of irrigation. He confirms the standard finding of constant returns to scale for inputs used. Ahmed also finds that owner-operated farms have higher productivity than those operated by tenants, thus lending support to the theory that tenancy arrangements such as share-cropping reduce efficiency in resource allocation. Moreover, Ahmed's results throw new light on the relationship of technological change to farm size and productivity. The smaller farms tend to cultivate (and double-crop) a higher percentage of their land with high-yielding varieties (HYVs) than do larger farms. However, the inverse relationship between farm size and productivity appears to be considerably weaker among those farms using HYV than among those still using traditional varieties, although there is no evidence that the Green Revolution has reversed the size productivity pattern from an inverse to a positive relationship. The technocrats' implicit assumption of the technological superiority of large farms is unsupported by the evidence. At the same time, the findings point to a clear tendency for smaller farms to be more labour intensive. Thus, in the case of Bangladesh, land reform or other small-farm strategies would tend to increase employment and distributional equity. If the farm size productivity relationships found on farms using traditional varieties are considered to be most representative of long-run patterns, these policies would tend to increase production. If, instead, the patterns currently observed on farms using HYV are considered representative of long-term trends, small-farm strategies would at least not reduce productive potential - although these policies might not raise output by as much as previous studies of the inverse relationship between farm size and productivity might have suggested. Even in this case, the equity aspects of small-farm strategies would make them an attractive option.

If national authorities eventually take note of the international body of empirical evidence on the farm size productivity relationship, ably augmented by this study on Bangladesh, and if they begin to redesign agricultural development plans accordingly, the prospects will brighten for agricultural development and for the achievement of more equitable income distribution in the 1980s, despite the growing constraints imposed by an increasingly less favourable international economic environment.

Washington DC
November 1980

William R. Cline
Senior Fellow
The Brookings Institution

Contents

LIST OF TABLES

AUTHOR'S ACKNOWLEDGEMENTS

This study reflects the efforts of a number of individuals who have contributed in various ways at different stages for the work.   To begin with, the generation of the data for this study could not have been possible without the contribution of a dedicated team of enumerators who helped me in directly implementing the survey of rural households in three Bangladesh villages. The tedious work of processing of primary data and the complex econometric analysis could not have been undertaken without the valuable assistance of David Viry and Josiane Capt in the use of the computer.   Pierre Antonilez helped me to apply the logit technique of analysis to some of the important parts of the study. Keith Griffin, A. R. Khan, David Freedman, Ajit Ghosh, Lothar Richter, Iqbal Ahmed, Jan Versluis, Amarjit Oberai and Bruce F. Johnston provided useful comments.   I am particularly grateful to William R. Cline for his advice and assistance in substantially improving and strengthening the analysis and in articulating the presentation of the results.

I am greatly indebted to Ajit Bhalla without whose continuous encouragement and guidance this research could not have been completed.

Last but not the least, I am very thankful to Irma King, Milli Kvistad, Francesca Rosner and Morwenna Lloyd for the long hours of typing of this manuscript and the large number of tables contained in it.

Glossary of Local Terms

Aus, Aman and Boro	are the three seasonal rice crops. They are respectively harvested in mid-summer, late autumn and spring. Aman is the main rainy season (July-November) and Boro is the dry season (December-April
Maund (abbreviation md.) and Seer (abbreviation sr.)	is the measure of weight and one maund is equivalent to 82.29 pounds or 37.33 Kg. Forty seers makes one maund (1 seer = 2.06 pounds or .933 Kg).
Taka (abbreviation Tk.)	is the unit of currency of Bangladesh. Currently about Tk. 15 exchange officially for one US dollar.
Acre	is the measure of area and is equivalent to .4047 hectares.
District	Administratively Bangladesh is divided into 4 Divisions which are further divided into 19 Districts which in turn are divided into 434 Thanas. An average Thana is about 125 square miles in area and has about 180 thousand people.

Chapter 1

Introduction

The Agrarian Scene

Incidence of rural poverty in Bangladesh has reached
dreadful proportions. While only 5 per cent of the country's
rural population could be categorised as extremely poor (those
in a state of acute malnutrition for much of the year) in 1963-
64, the proportion rose to over 40 per cent in the 1970s [42].
The priority question for planners in Bangladesh is how to increase productive employment and incomes of the rural poor (those
severely undernourished) who form over 60 per cent of the rural
population.

Rural development in Bangladesh is virtually synonymous with
agricultural development. Crop production is estimated to have
contributed nearly 6C per cent to the conventionally measured
GDP in recent years. Furthermore, Bangladesh agriculture has
the greatest dependence on a single crop, rice. Except Vietnam,
no other country in South and South-East Asia has even half as
much of gross cropped land under rice as a percentage of net
cultivable area. It is little wonder, that the growth in Bangladesh
agriculture over the last two decades relied heavily on the
growth of rice production. In 1975 rice output per hectare in
Bangladesh was about the same as in India and Thailand and only
marginally higher than that in the Philippines. It was lower
than in all other Asian countries and no more than three-quarters
of the Asian average [44]. About three-quarters of all employment
in crop production in Bangladesh was contributed by rice cultivation. Thirty-seven per cent of the available man-days in
agriculture remain unutilised. Therefore, in view of such an
overwhelming quantitative importance of the rice crop, a
significant increase in labour input in its cultivation will have
a major impact on overall rural employment. Similarly, a rise
in productivity of rice cultivation would be reflected in a
significant growth in agricultural production.

By the standards of the South and East Asian countries
Bangladesh agriculture is characterised by the most acute
land scarcity, the highest labour abundance, the greatest
dependence on rice and about the lowest yield of rice per
acre.   Although the agrarian scene in Bangladesh reveals high
population densities and generally small holdings, land ownership distribution is quite unequal.   In 1977, 3 per cent of
the households were found to own more than 25 per cent and 11
per cent owned more than 52 per cent of all land.   In contrast,
one-third of rural households owned no agricultural land at all
and, together with those owning less than half an acre, the
effectively landless constituted 48 per cent of the rural population [62].   About 30 to 35 per cent of the farms are involved
in share-cropping arrangements under which the tenants generally
bear all the input costs, but share 50 per cent of the output with
the landlord.

Objectives of the Study

An ambitious target of doubling food crop production within
the second five year plan period (1980-85) has been adopted by
the Government of Bangladesh [31].   This is to be achieved primarily through greater utilisation of land and extending irrigation
facilities to cover the whole country.   The strategy implies the
provision of a package of modern agricultural inputs including
new high yielding variety (HYV) seeds, fertilisers and controllled
water supply (low-lift irrigation pumps and tubewells).   In the
past, priority was given to the larger and richer farmers in
the allocation and chanelling of this package of heavily subsidised
modern inputs based on the notion that they have superior yields
to those of the "nonprogressive" farmers.[1]   This monograph examines

---

[1]Subsidy as a percentage of total cost were 68 per cent and
77 per cent for low-lift pump and deep tubewell respectively [33].
In 1977-78 the subsidy as percentage of cost was 48 per cent for
urea (nitrogen) fertiliser, 67 per cent for phosphate (TSP)
fertiliser and 60 per cent for potash (MP) fertiliser [34].

the validity of this thesis. The two related and specific
questions addressed in this study are the following. (a) Do
large farms tend to be more or less efficient than small farms? and
(b) Has the Green Revolution (popular name of the technology associated
with HYVs) altered this relationship between farm size and productivity? The latter question is of considerable policy interest
not only for Bangladesh but for the entire range of Asian countries
where the diffusion of the HYV technology has been greater and
more widespread. No clear cut empirical evidence appears to be
available to shed light on this important question. The few
observations that have been made are speculative and conjectural
rather than being based on any rigorous and systematic treatment
of sufficiently disaggregated data [13]. Therefore this study
will attempt to fill this significant gap in empirical work on
this controversial but important issue. For an overall assessment
of social efficiency across farms where land is the most scarce factor,
the relationship between farm size and the degree of land utilisation (i.e. land double-or multiple-cropped) is considered.

An attempt is made in this monograph to examine the relationship between farm size and the intensity of labour use. It will
examine whether the advent of the Green Revolution technology
changes this relationship. Since emphasis is being placed on
mechanised cultivation by the planners in Bangladesh (see, for
example, Guidelines on the Second Five Year Plan: 1980-85), the
analysis of disaggregate operation - wise labour-use data in crop
production is undertaken to determine whether all types of mechanisation (such as tubewells for irrigation versus tractors for
tillage) are labour-displacing. Given the fact that the number of
landless labourers has been increasing both absolutely and as a proportion of the agricultural population, the study examines in-
depth the technical and institutional factors which determine the
demand for hired labour.

The monograph covers another important institutional factor,
land tenancy. The specific question being examined is whether the
existing tenurial arrangement tends to depress productivity and

labour-use on account of the tenurial insecurity as well as the
prevailing crop-sharing system. If this tendency is observed in
a traditional agricultural setting, is it likely to persist under
conditions of technological change? The relationship between
tenancy and cropping intensity is also examined.

It is by now well known that the existence of factor market
imperfections in the rural areas result in much easier access to
factors of production for some groups than others. For example,
the political power and social status associated with ownership
of land enable large land owners to exert the most influence over
sources of institutional credit and government agencies responsible
for the distribution of subsidised modern inputs (such as irrigation
water, fertilizers, etc.) and extension services (knowledge of
the new technology). This study examines the nature and degree
of such rural factor market imperfections and evaluates the effectiveness of the various specialised agencies (such as Bangladesh
Agricultural Development Corporation, Integrated Rural Development
Programme, etc.) established by the government in supplying credit
and key agricultural inputs and extension services to the less
privileged sections of the village community such as the small
and marginal farmers and agricultural tenants.

Finally, the Green Revolution technology which requires the
use of HYV seeds, controlled doses of chemical fertilisers and
irrigation water is supposed to be technically scale-neutral
and thus within the reach of all categories of farmers. Therefore,
this study explores whether any relationship exists between the
prevailing agrarian structure (institutional factors like farm-size
distribution, tenancy and membership in cooperative societies) and
adoption of HYV innovations. In addition, the study examines
the influence of (a) demographic factors like age of cultivator
and the number of adult agricultural workers in the household and
(b) the cultivator's level of education on the adoption of HYV.

In sum, the principal objective of this monograph is to
contribute to the analytical and empirical basis for answering the
policy question of whether and to what extent changes in the prevailing agrarian structure could successfully increase agricultural

production and improve rural employment and income distribution
in Bangladesh both under a traditional agricultural setting and
under dynamic conditions of technological change. However, the
analysis in the study is technical in nature which points to the
potential of certain strategies and it does not attempt to spell
out a blue print of action for agrarian reform containing the
specific details of its implementation from the administrative
and political perspectives.

Source of Data and Basic Characteristics
of Survey Area

The data are based on the complete enumeration of 459 rice
holdings in three villages in Bangladesh, one each in the districts
of Bogra, Sylhet and Noakhali for the 1975 Aman and 1974-75 Boro seasons. Information was collected separately for farms using local
varieties and those using HYVs. Over 76% of these 459 households
had grown HYV rice during Aman and/or Boro seasons under
reference. In Bogra, Noakhali and Sylhet villages these figures
were respectively 50%, 99% and 84%. On an average the area under
HYV accounted for 26%, 74% and 57% of the total Boro rice area
respectively in the Bogra, Noakhali and Sylhet villages. The
Bangladesh national average for Boro 1976-77 was 58%. Data on
the use of fertilizer and organic manure were also collected for
individual farms. About 60% of the 459 farms reported the use
of irrigation water and this information was collected by the
method of irrigation adopted. The 305 farms using irrigation
reported that 55% of the total area irrigated was by power pumps
average of 55% (compared to Bangladesh); 6% by hand-tube-wells
(Bangladesh average being less than 1%) and another 8% by means of
powered tube-wells (Bangladesh average for this is 6%); and 38%
by canal and indegeneous methods (Bangladesh average being 41%).[1]
Separate information was also obtained for farms using bullocks and
those using mechanised ploughing (i.e. using tractors and power
tillers). Of the 459 holdings, 60 in Aman season and 63 in the Boro

---

[1]The national figures have been obtained from [62].

season used mechanised ploughing; 40 of these were located in
the Noakhali village alone. Cropping intensity in Bogra (220%)
and Noakhali (188%) were much higher than the national average
(148.5%) and that in the Sylhet village (126%) was the lowest.

Turning to institutional factors, land ownership distribution was found to be more unequal in the three villages compared to the national situation when viewed in terms of the Gini
coefficient of land ownership distribution. It was .68 for the
three villages combined as against .57 for the whole of Bangladesh
(see Table 1). However, landlessness is more acute in the
survey villages compared to the overall national situation. Between
one-third (Sylhet) and one-quarter (Bogra) of the households did
not own any land. Compared to this about 11% of the rural households
of entire Bangladesh were found to be landless. Proportion of
total rural households, landless or nearly landless, (owning
none or less than one acre of land) were 71%, 53% and 68%
respectively in the Noakhali, Bogra and Sylhet villages. This
proportion was 58.5% for all Bangladesh [32]. Average farm size
in the survey area is 1.84 acres which is less than the national
average. The size in Noakhali, Bogra and Sylhet villages were
respectively 1.08, 1.56 and 2.45 acres. The coefficient of
variation for the three villages were respectively 1.1, 0.83 and
1.1.

Incidence of tenancy in these three villages combined is not
much different from the overall national situation; although there
is a wide variation among the three villages(Table 2).In Bogra much less
than one-half of the farms were owner farms as against nearly three-
quarters in Noakhali. The Bangladesh average is 61%. Like the
rest of Bangladesh tenants pay 50% of their output to landlords as
rent. About a quarter of the land operated in the survey villages
is leased-in implying that tenant farming, a necessary sign of
"feudalism", is quantitatively and relatively still quite important.[1]

Participation in farmer organisations was the highest in
Noakhali (68% of the households) and lowest in Sylhet (26% of
households) with Bogra (42% of households) in between. Nearly
60% of the 304 members of farmer's organisations belonged to

---

[1] For an elaboration of this point, see [2].

the Comilla-type village cooperatives (KSS) and other multi-purpose cooperative societies. Another 40% of the members of farmers' organisations belonged to irrigation groups.

Design and Some Highlights of the Study

A brief description of the theoretical and conceptual framework is presented and the nature of hypotheses being tested are discussed at the beginning of each chapter.[1] The methodology applied in the analysis of the cross-sectional farm-level data, also elaborated, varies between chapters. For example, in Chapter 2, use is made of multivariate (logit and multiple regression) analysis to examine the influence of the prevailing agrarian structure, education and demographic variables on the adoption of HYV (using several types of indices of adoption). In the following chapter (Chapter 3), it will be seen that the returns to scale in Bangladesh agriculture are not significantly different from constant returns and this finding is equally valid for the traditional and the high yielding variety (HYV) rice crops. What is more interesting is that, under both a traditional agricultural setting and under conditions of technological change, an _inverse_ relationship between land productivity and farm size has been observed in conjunction with constant returns to scale in Bangladesh agriculture. Prevailing tenurial system appears to depress output in the cultivation of both traditional and HYV crops. These conclusions were found to be valid even after controlling for exogeneous land quality (soil fertility) and endogeneous (man-made) land improvement factors. Although, the above relationships between farm productivity and agrarian structure does not appear to be a fully universal law, this feature occurs repeatedly in different combination of crop varieties and seasons in different parts of the country. A significant _negative_ relationship between cropping intensity and farm size is observed even after controlling for the availability (participation in irrigation pumps groups) and use (percentage area irrigated) of irrigation water. One important reason for this could be the big farmers remaining engaged in a multiplicity of

---

[1] The various hypotheses are derived from existing theory or are based on a _priori_ reasoning and often reflects evidence from other countries.

channels of profit making and the cultivation of land appears on
their agenda only for a brief period; at other times of the year, thei
land remains fallow.

Chapter 4 examines the influence of farm size and tenancy on
labour utilisation after controlling for endogeneous technical
factors like mechanisation, irrigation method used, fertiliser-
use and improved agricultural practices (frequency of weeding, row
planting, transplanting, etc.). Multiple regression analysis of
farm-level cross-sectional variatons in labour intensity (total and
operation-wise) per unit of land was undertaken for both HYV and
traditional crops. A picture similar to the relationship between
agrarian structure and productivity emerges. In the context of an
acute and growing level of landlessness, the determinants of the
demand for hired labour were examined in relation to both the
agrarian structure and technological change.

The existence and implications of the imperfections in the
rural factor market (credit and fertiliser) is analysed in the
context of the green revolution technology in Chapter 5. The
issue of unequal access to knowledge on HYV and to farmers'
organisations is also taken up.

Table 1
Size Distribution of Total Land Owned by
Rural Households, Bangladesh: 1976-77
(per cent of total households)

Size Region of total land owned (acres)	Bangladesh[a] 1977	Three Villages 1976	Bogra 1976	Noakhali 1976	Sylhet 1976
Zero	11.07	27	24	24	31
0.01 - 1.00	47.44	35	44	47	22
1.01 - 2.00	16.43	16	18	16	16
2.01 - 3.00	8.91	8	8	5	10
3.01 - 4.00	5.27	5	3	5	7
4.01 - 5.00	3.29	2	0	1	4
5.01 and above	5.50	5	3	1	9
Gini coefficient	.57[b]	.68	.71	.68	.66

[a]Source: [32]
[b]Source: [38]

Table 2
Incidence of Tenancy in Bangladesh 1976-77

Region	Year	Owner Farms	Part-tenant and Part-owner	Tenant Farms
		in % of all farms		
Bangladesh[a]	1977	61	32	19
Three Villages Combined:	1976	53	36	11
Bogra	1976	45	48	7
Noakhali	1976	73	21	6
Sylhet	1976	50	35	16

[a]Source: [38]

Chapter 2

Agrarian Structure and the Adoption of HYV Technology

Introduction

In this chapter the relationship between the prevailing agrarian structure and the adoption of HYV technology in Bangladesh is examined. More specifically, the influence of institutional factors like farm size, tenancy and participation in farmers organisations on the adoption of HYV is investigated. In addition, the influence of demographic variables like the age of the cultivator and the number of adult agricultural workers in the household on HYV adoption is assessed. It is also investigated whether there exists any relationship between the level of education and adoption of HYV innovations. Finally, the relationship between the prevailing agrarian structure and the dynamics of HYV adoption is reviewed. Before embarking on an empirical verification of these relationships with the help of bivariate and multivariate analyses of the cross-sectional survey data, a brief description of the theoretical and conceptual framework is presented and the nature of the hypotheses being tested are discussed.

Indices of HYV Adoption

A farmer is defined to have adopted HYV if he has placed any part of his rice acreage under the new varieties. The adoption rate is expressed as a ratio between the number of farms using HYV seeds and the total number of farms. No less important than the adoption rate is the 'intensity of adoption' which measures the proportion of rice area (instead of cultivators) brought under HYV. One other measure of HYV adoption which combines the crude adoption rate and the intensity of adoption is the index of participation which is obtained simply by multiplying the value of one ratio

by the other. One final measure of HYV adoption is the
'propensity to adopt HYV' which permits us to measure the
likelihood of a cultivator adopting HYV through the use of
a special multivariate statistical technique (logit analysis
as described later in the text and in the appendix), and
this also allows us to rank a set of explanatory variables
in terms of their relative influence on a farmer's propensity
to adopt.

Farm Size and HYV Adoption

It is by now fairly well established that the rate of
HYV adoption is usually higher among larger farmers.[1]   A
number of explanations have been put forward for the lower
adoption rates observed for the smaller farms. The first major
reason is the small farmer's attitude to risk
in changing to the new seeds and cultural practices. The
cultivation of HYV brings with it new inputs whose rate and
timing of application and various combinations over the cropping
season are highly complex and calls for the use of improved
cultural practices. Thus they are generally uncertain about
their returns from land in such a situation. Indeed in
Bangladesh variability of per acre HYV rice output is considerably
higher than that of the traditional rice varieties particularly
during the Aman season (Table 3). HYV rice is certainly a
riskier crop during the Aman season (coefficient of variation
of HYV rice output is .41 compared to .39 for the traditional
rice variety). In addition, considerable weather risks associated
with the cultivation of rice almost wholly dependent on rainfall

---

[1]For empirical evidence on this drawn from a number of
regions of India, see [23, 26, 35,49 and 59]. This is also
supported by evidence from some parts of Bangladesh [9],
Indonesia [53] and Malaysia [14].

during the main season (Aman) for rice cultivation, particularly
during the critical periods of sowing and flowering of the
rice plant.[1] Since adopters of HYV during the Boro season
grow their rice primarily on irrigated land, their risk is
minimised considerably (coefficient of variation of HYV Boro
output is the lowest of all crops).

While the risk and uncertainty attached to HYV cultivation
exists for farmers of all sizes, the degree of uncertainty is
higher for the smaller farms. It is partly because access
to knowledge of the use of the new technology is imperfectly
distributed between the small and large farmers. As will be
seen in Chapter 5, while the bulk of the small farmers in
Bangladesh rely on indigenous sources of information on the
new technology, a higher proportion of the larger farmers have
greater access to government agricultural extension agencies.

The small farmer is also confronted with market uncertainty
arising from variation in the price of output and inputs. Given
the high costs of HYV cultivation and the market orientation of
its production, it is inevitable that the decision to adopt the
new seeds is sensitive to market prices.[2] Though these uncertainties may be directly relevant to the surplus farmers, these

---

[1]
  The coefficient of variation of rainfall in inches is
.52 during the flowering stage of the rice plant and .25
during the sowing/transplanting operation [8].

[2]
  The HYV adopters marketed over one-tenth of their Aman
rice output as compared to about one-twentieth by the nonadopters
(this difference is statistically significant at the .0005
level).   During the Boro season, HYV adopters marketed nearly a
fifth of their rice output as compared to less than one-tenth
by the nonadopters (this difference is statistically significant
at .0005 level).

are important for relative profitabilities of different crops
and cropping patterns in the Bangladesh content [25]. In general, higher
uncertainty exists for the smaller farms also because they are
likely to get a lower price for their produce compared to their
larger neighbours because of the formers' lack of storage
facilities and inability to hoard the grain until the price is
favourable. In the input market, the cost and supply of
fertiliser, irrigation, water, etc. introduces uncertainties.
As will be seen in Chapter 5, while the unreliability of the
supply of fertilisers affects the entire farming community
(89% of the 290 fertiliser-users in the survey area obtained
part of their supplies from the black market and on an average
this source accounted for 63% of the total fertilisers purchased),
the worst sufferers were the small farmers (the proportion of
fertiliser purchased in the black market is negatively correlated
with farm size). In so far as fertiliser serves as an important complementary
input, this phenomenon will be reflected in the HYV adoption rate.
    The small farmer's greatest disadvantage in relation to the
large farmer is in terms of his risk bearing capacity. Given
their limited resources, the small farmers are forced to
operate with a short time horizon which restricts the scope
for adopting the new technology. Where a large farmer with the
security of a large stock of assets is capable of planning
over a period of several years, and of withstanding occasional
losses from crop failure, a small farm can plan only on a year
to year basis. Indeed the large farmers in Bangladesh are
found to be engaged in a multiplicity of channels of profit
making (as elaborated in Chapter 3) and gross income increases
significantly with increase in farm size.[1] Moreover, a
significant positive relationship between farm size and ownership of farm assets is observed in the survey area. The
implication of all this is that the large farmer has the
ability to make the necessary investment in inputs either from

---

[1] Correlation coefficient between gross income and farm
size in the survey area is found to be +.36.

his own resources or by borrowing. The large farmer
has a high credit-worthiness rating and has easier access to
sources of cheaper institutional credit (see chapter 5 for
evidence from survey area)[1].

Some have argued that the intensity of adoption, i.e.
the proportion of rice area under HYV is a better measure of
the adoption of the new technology. In contrast to the
positive relationship between farm size and the crude adoption
rate, an inverse relationship has been observed between farm size
and the intensity of adoption.[2] Uncertainty is the major
constraint to adoption among small farmers, but once the fear
is overcome and they decide to adopt, the latter prefer to
put in as much an effort as possible for HYV cultivation.
Furthermore, given the high overhead cost of the decision
to adopt - in gathering information about the performance
and technology of the new varieties and in procuring the
necessary inputs and credit - in relation to the size of his
holding, a small farmer is likely to be more committed to the
new variety in terms of the proportion of acreage under HYVs than
the larger farmers.[3] It has also been argued that given various
supply constraints on inputs, larger farms would prefer to
concentrate the riskier crop on one part of their land, thereby
diversifying his risks, rather than spreading them over the
whole land area.

In India, farm size appears to be positively related to the
index of participation, the third measure of HYV adoption [49].
In other words, the positive relationship between farm size and
crude adoption rate outweighs the negative relationship between
size and intensity of adoption.

While the likely nature of the static cross-sectional
differences in HYV adoption pattern across farm size is discussed
above, not enough evidence has been generated to shed light on
the dynamics of this relationship. A set of evidence from

---

[1] Similar evidence is observed for other areas of Bangladesh [9].
[2] For evidence from India see [49,59] and from Bangladesh [8].
[3] More fundamentally, the pattern is probably another manifestation of
the general land use intensity/farm size relationship. Productivity
per acre being higher on small farms, one way of reaching higher productivity levels could be to plant a higher proportion of cultivated
land under HYV.

the Far East suggests that the large farmers adoption rates
tended to be higher in the long run, though not so initially
[53]. On the other hand, evidence from Pakistan reveals a
general tendency for larger farms to adopt HYV wheat earlier
than the small farms, but for the lag to be short [50].

Tenancy and HYV Adoption

In addition to the type of risk and uncertainty discussed
above, which affect the return from the land, the share-
croppers are confronted with the additional uncertainty of
tenurial insecurity. Under the prevailing share-cropping
system where no recorded contract between the landlord and
the tenant exists so that the latter may be evicted at will
by the farmer. Although share-cropping system divides risk
between the landlord and the tenant, sharing of output means
that a tenant with a certain output will have less for subsistance than an otherwise equally placed owner-farmer. It
has also been argued that the tenant farmer will be more
desperate than an owner farmer in adopting the riskier crop.
Cross-country evidence is mixed. Evidence from parts of
India [23] and Indonesia [53] reveals higher adoption rates
achieved by tenant farmers as compared to **owners**. Another
set of evidence from Sri Lanka [27], Malaysia [14] and
Philippines [48] indicates no definite pattern in HYV adoption
with respect to a change in tenure status among cultivators.
On the other hand, evidence from one part of India [56] and
Indonesia [54] suggests higher adoption rates for owner-operators
compared to that of tenants.

Membership in Farmer Organisation and HYV Adoption

Adoption rates are observed to be higher for members of
farmers' organisations.[1] Membership in farmers' organisations
brings with it numerous advantages. Risks and uncertainty is
reduced through improved access to sources of knowledge and
information on the new technology. Uncertainties with respect

---

[1] Evidence available for Malaysia [14], India [56] and
Bangladesh [4 and 57] suggest this relationship.

to supply of key inputs such as fertilisers, irrigation water, etc. are reduced. Equally important is the improved access to cheaper sources of credit.

However, when one looks at the dynamics of HYV adoption in one area of Bangladesh, it is observed that while members of farmers' organisations pioneered HYV adoption, the non-members caught up with the member's adoption rates within a remarkably short time [28].

Demographic Variables and HYV Adoption

In one region of India the innovative farmers were founds to be young in age [17] while in another area age did not appear to be a good predictor of HYV adoption [56].

Number of adult agricultural workers in the household in as much as it eases the labour constraint is believed to influence farmer's decision to adopt HYV.

Education and HYV Adoption

Adoption rate is believed to have a positive relationship with the level of education. Education is assumed to have an innovative, allocative and worker effect on farmers as well as an externality [21]. The innovative effects is argued to consist of (a) the ability to derive new information - know what, why, where, when and how; (b) the ability to evaluate costs and benefits of alternative sources of economically useful information; and (c) the ability to quickly establish access to newly available, economically useful information. The allocative effect would consist of the ability to choose optimum combinations of crops, new inputs and agricultural practices quickly covering both business and production activities. The worker effect would consist of improvements in the quality of labour such as ploughing and harvesting skills. The externality arises from the fact that neighbouring farmers and other producers in the vicinity who are in direct contact with educated farmers would be able to consult the educated farmers without paying and would be able to copy (without paying) his sources of information, crop and input combinations

and related production and business techniques of proven
success.

Hypotheses

To sum up, the hypotheses that are being tested can be
listed as follows. The crude adoption rate (proportion of
farmers adopting HYV) is expected to be <u>positively</u> associated
with farm size; it is expected to be higher for members of
farmers' organisation; it is expected to be **negatively**
associated with the tenurial status of the cultivators. The
propensity to adopt HYV is expected to be <u>greater</u> for larger
farms, for owner-operators (as compared to tenants), for
members of farmers' organisations (compared to nonmembers),
for cultivators having larger number of adult agricultural workers
in the household and for farmers with higher levels of education.
Intensity of adoption is expected to be <u>negatively</u> associated
with farm size and tenancy; it is expected to be <u>positively</u>
associated with membership in farmer organisation, percentage
rice area irrigated, number of adult agricultural workers in
the household and with the level of education. The third
measure of HYV adoption, index of participation, is expected
to be <u>positively</u> associated with farm size, membership in
farmer organisation; it is expected to be <u>negatively</u> associated
with tenancy.

No assumption is being made regarding the relationship
between the dynamics of HYV adoption and the agrarian structure.[1]

Empirical Verification

In the following section we proceed with the empirical
verification of these hypotheses by analysing farm-level cross-
sectional variations in HYV adoption using data for both the
Aman and Boro seasons seperately and combined. Variations
in the crude adoption rate, the index of participation and
the dynamics of HYV adoption are analysed with the help of
bivariate tables. Inter-farm cross-sectional variations in

---

[1]In general one observes an S-shaped adoption curve over
time.

the propensity to adopt HYV is analysed with the help of
special type of multivariate analysis known as the logit
analysis which is applied   when the dependent variable is
binary (See appendix for elaboration). Interfarm cross-
sectional variations in the intensity of HYV adoption is
analysed with the help of ordinary least squares multiple
regression analysis. We begin with an analysis of the level
and determinants of the crude HYV adoption rate.

Proportion of Farmers Adopting HYV and Agrarian Structure

About 44% of the 459 rice farms adopted HYV during the
Aman season as compared to about 66% during the Boro season.
This reflects the seasonal differences in the crude adoption
rate. As expected, a significant <u>positive</u> relationship between
the adoption rate and farm size is observed (Table 7). Only
during the Aman season this relationship is found to be weak.
A significant <u>difference</u> is observed in the adoption rates with
a change in the tenure status of cultivators with owner cultivators revealing the highest adoption rates during the Aman
and Boro seasons seperately and in combination (Table 8). The
tenant farms show the lowest adoption rates during Aman and
when the two seasons are considered in combination. The rate
of adoption by owner-cum-tenant farms lies intermediate between
the owner and tenant farms (the Boro season being the only
exception). Adoption rates are significantly <u>higher</u> among
members of farmers' organisations during all seasons (Table 9).[1]

Propensity to Adopt HYV

Having noted the relationship between the crude adoption
rate and the prevailing agrarian structure, by means of bivariate
tables, a multivariate analysis is now undertaken to explain

---

[1]Evidence from other regions of Bangladesh supports this
finding. The 228 farmers who had adopted HYV Aman in 1970 in
7 districts were all members of cooperatives with varying degrees
of cooperative experience [57]. Similarly, evidence from Rajshahi
and Rangpur districts reveal that during 1972-73, 64% of the
296 member farmers adopted the new technology as compared to
only 36% of the 117 nonmembers [4].

the probability of an individual cultivator becoming an
adopter of HYV. Under this approach the dependent variable
for each observation takes a value of one if the respondent
has adopted HYV, while a value of 0 is assigned to those individual cultivators who have not adopted HYV. The dependent
variable being dichotomous, the appropriate type of multivariate analysis which we are using is logit analysis (a
description of this statistical technique is presented in the
appendix). The results are presented in Tables 4 and 5. While
Table 4 presents results based on a mixture of continuous and
binary explanatory variables, results based exclusively on
binary explanatory variables are presented in Table 5. The
latter table permits us to rank the individual explanatory
variables in terms of their relative contributions in explaining the variations of the dependent variable.

As expected, propensity to adopt HYV is _positively_
associated with farm size and membership in farmer organisation;
it is _negatively_ associated with tenancy and these associations
are observed for both, the Aman and Boro seasons seperately
and in combination (Table 4). From Table 5 we can conclude
that (a) large farmers (farm size 5.5 acres and above) are
more likely to adopt HYV than small farmers (size less than
2.5 acres); (b) medium farmers (with size range 2.5 - 5.5 acres)
are more likely to adopt HYV compared to small farmers;
(c) owner-operators are more likely to adopt HYV compared to
tenant farmers (no significant difference is observed in the
propensity to adopt between owner-cum-tenants and pure tenants);
and (d) members of farmers organisations are more likely to
adopt HYV compared to nonmembers. When these explanatory binary
variables are ranked in terms of their explanatory power, membership in farmers' organisations always ranks first. Large farmers,
medium farms and owner farms have ranked in different orders
as the seasons vary and when the two seasons are combined (Table 5).

Age of the cultivator has no relationship to his propensity
to adopt HYV (Table 4). Number of adult agricultural workers
in the household also has no relationship to the propensity to

adopt HYV except for the Boro season and the two seasons combined when regional influence is not controlled (it is found to be negative). This indicates that availability of labour does not act as a constraint on the propensity to adopt HYV. Educational level of the farmer does not appear to have any relationship with the propensity to adopt HYV; there is one exception during the Aman season. Farmers who have completed their education upto the primary level and above are more likely to adopt HYV during Aman when compared to farmers who are illiterate. It has already been noted that risk and uncertainty with respect to the output from the land is the highest for HYV Aman. In such a situation the farmers derive benefit from the primary and higher level of education in their decision to adopt. This finding is confirmed by similar evidence from Northern India. It has been observed that just literacy and lower primary education among cultivators in the situation represented by the Green Revolution in Punjab and Haryana do not explain the diffusion of technology, suggesting that in a dynamic agriculture, with a technology of the Green Revolution type, mere literacy is not enough. Sustained rural education up to secondary level is required [20].

Finally, regional variations in the propensity to adopt are observed. During both the seasons, farmers in Noakhali are more likely to adopt HYV as compared to those in Bogra and Sylhet (Tables 4 and 5).

Intensity of HYV Adoption

On an average adopters cultivate HYV on nearly 70% of their rice area during Boro as compared to 38% during Aman (Table 6). This reflects the wide seasonal differences in the intensity of HYV adoption. This is little surprising because of the higher risks in respect of HYV output during the Aman season as compared to the Boro (Table 4 and earlier discussions).

During the Aman season intensity of HYV adoption is
negatively associated with farm size and tenancy; it is
positively associated with membership in farmers' organisation
(Table 6). When regional differences are taken into account
these relationships are no longer significant. Similarily,
during the Boro season, intensity of HYV adoption is negatively
associated with farm size; it is positively associated with
membership in farmers' organisation. No significant relationship is observed with tenancy. When regional differences are
taken into account, these relationships during Boro are no
longer significant.Intensity of adoption does not appear to have any
significant relationship to number of workers in the household (except
during Aman without regional dummy).
Intensity of HYV adoption does not appear to have any
relationship with the educational level of the cultivator.
However, as was observed with respect to farmer's propensity to
adopt during Aman, the only educational variable influencing
the intensity of adoption is education up to primary and
above (when regional differences are taken into account,
this is no longer significant).

Significant regional differences in the intensity of
HYV adoption is observed. It is lower in Sylhet compared to
Noakhali. While it is lower for Bogra farms compared to
those in Noakhali during Aman, it is significantly higher during
Boro.

Index of Participation

When, the crude adoption rate is combined with the intensity
of adoption to produce an index of participation, no definite
relationship between farm size and index of participation is
observed (Table 7). This implies that the positive relationship
between size and crude adoption rate is neutralised by the
negative relationship between size and intensity of adoption.

It appears that the higher crude adoption rate among owner-
cultivators is reinforced by their higher intensity of adoption
to produce a higher index of participation among owner-cultivators
compared to the tenants (Table 8). Moreover, the owner-cultivators
are observed to have a higher index of participation compared to

owner-cum-tenant farms. During Aman the owner-cum-tenant
category of farms also have a higher index of participation
compared to that of the tenant farms. However, during Boro,
the higher intensity of adoption among tenant farms compared
to owner-cum-tenant farms results in higher index of participation for the tenants because there is not much difference
in the crude adoption rates of these two farm categories
during that season.

Members of farms' organisations have both significantly
higher adoption rates and intensity of adoption and this
has resulted in considerably higher index of participation
among member farms.

Dynamics of HYV Adoption

It appears that the crude adoption rates do not initially
differ across farm size, although the adoption rates among
large farmers tend to be higher in the long run (Table 10).
Although this gap in adoption rate across farm size has emerged
over time, it is to be noted that the smaller farms have also
attained high adoption rates over time. One reason for nonadoption by smaller farms could be the high cost of cultivation
involved. If small farms lagged behind the large farms in adoption
rates due to the high costs of cultivation and due to their
inability to make the necessary investments in HYV inputs,
either from their own resources or by borrowing, it is argued
that this gap will continue indefinitely in the absence of
policy intervention. On the other hand, if the reason for
nonadoption is the uncertainty attached to the cultivation of
the new varieties, the differences may diminish to some extent
over time, without policy intervention as experience with
the new crop and increased knowledge about cultural practices
reduce risk and uncertainty.

Although the owner-cultivators took the lead in HYV
adoption, in the long run, owner-cum-tenant cultivators caught
up with them (Table 11). The tenant cultivators adoption
rates were initially the lowest and they continued to be
laggards over time. Insecurity of tenancy and the prevailing

share-cropping system continues to depress their rates of
HYV adoption over time.

Members of farmers' organisations pioneer HYV adoption
and this lead over the nonmember cultivators is maintained
over time (Table 12). Since the nonmembers did not adopt
the new varieties due to their reduced or lack of access to
financial and material resources, this difference is expected
to continue for an indefinite period in the absence of
appropriate policy intervention. However, the fact that
the nonmembers adoption rates have increased over time and
the gap diminishes to some extent indicates that risk and
uncertainty are reduced even in the absence of policy intervention due to the members' acquisition of experience with
the new crop and increased knowledge about the cultivation
methods.[1]

Summary and Conclusions

During the Aman season both the adoption rate (44% of
farms) and the intensity of adoption (30% of rice area under
HYV) are found to be considerably lower than those during
the Boro season (66% of farms place 70% of their rice area
under Boro HYV). This reflects the fact that the risk and
uncertainty associated with the cultivation of HYV Aman
crop is higher than that of the HYV Boro crop (coefficient
of variation of HYV Aman rice output is .41 compared to .37
in the case of HYV Boro).

The bivariate analysis of the interfarm variations in the

---

[1]However, evidence from another area (Comilla) of
Bangladesh reveals that the nonmembers were initially laggards
but they did not trail behind for very long and their adoption
rates reached the level of member cultivators' adoption
rates [28].

crude adoption rates during the Aman and Boro seasons separately
and in combination shows that it is _positively_ associated with
farm size. Tenancy appears to have a _depressing_ effect on
adoption and members of farmers' organisations are observed
to have significantly _higher_ adoption rates compared to nonmembers. Using a multivariate (logit) analysis it is
observed that the propensity to adopt HYV is _positively_
associated with farm size and membership in farmers' organisations; it is _negatively_ associated with tenancy. These
associations are observed during both the Aman and Boro
seasons separately and in combination. It can be concluded
that both the large- and medium-sized farms are more likely
to adopt HYV compared to the small farms; owner operators
are more likely to adopt compared to the tenants; and members
of farmer organisations are more likely to adopt compared to
the nonmembers. When these binary explanatory variables are
ranked in order of their relative impact on the propensity
to adopt, membership in farmer organisations always ranks
first and the ranking of the remaining significant variables
follow a mixture of ordering which varies with seasons.
Propensity to adopt bears no significant relationship to either
of the two demographic variables (age and family structure)
implying that innovativeness of a farmer is not related to his
age and that on-farm labour availability is not a significant
constraint to HYV adoption.

Just literacy and lower primary education do not appear
to have any significant effect on either the farmer's propensity to adopt or on his intensity of adoption. However,
cultivators who have completed education upto primary level
and above are more likely to adopt HYV Aman compared to
illiterate farmers and the intensity of adoption is also
significantly higher for the former. It is interesting to
note that these differences have surfaced in the cultivation
of the riskiest crop.

Intensity of HYV adoption is _negatively_ associated with
farm size during both seasons; it is _negatively_ associated with
tenancy during the Aman season; and it is _positively_ associated

with membership in farmers' organisations during both seasons.
When regional differences are taken into account, these relationships are no longer found to be significant.

It is observed that the positive relationship between
size and the crude adoption rate is neutralised by the negative
relationship between size and the intensity of adoption with
the result that no definite pattern of relationship emerges
between size and index of participation. The higher crude
adoption rate among owner- cultivator is reinforced by their
higher intensity of adoption to produce a higher index of
participation among owner-cultivators compared to the tenants.
Moreover, the owner cultivators are observed to have a higher
index of participation compared to owner-cum-tenant farms.
During Aman the owner-cum-tenant category of farms also have
a higher index of participation compared to that of the
tenant farms. However, during Boro the higher intensity of
adoption among tenant farms compared to owner-cum-tenant
farms results in higher index of participation for the tenants
because there is not much difference in the crude adoption
rates of these two farm categories during that season.

The intensity criterion warrants the more weight as it
helps answer the question: do small or large farmers take more
advantage of HYV technology? It is clearly the small farmer who
do so in view of the fact that they plant a greater percentage of
rice area under HYV. No doubt a higher proportion of large farmers
adopt HYV, the crude adoption rate would be interesting to look at
only when intensity of adoption is constant per adopter.

Members of farmers' organisations have both significantly
higher adoption rates and intensity of adoption and this has
resulted in considerably higher index of participation among
member farms.

From the dynamics of HYV adoption it appears that the
crude adoption rates do not initially differ across farm size;
however, the adoption rates among large farmers tend to be
higher in the long run. Although this gap in adoption rate

across farm size has emerged over time, it is to be noted
that the smaller farms have also attained high adoption rates
over time. Although the owner-cultivators took the lead in
HYV adoption, in the long run owner-cum-tenant cultivators
caught up with them. The tenant cultivators' adoption rates
were initially the lowest and they continued to be laggards
over time. Insecurity of tenancy and the prevailing share-
cropping system continues to depress the tenant cultivators'
rates of HYV adoption over time. Members of farmers' organisations pioneer HYV adoption and this lead over the nonmember
cultivators is maintained over time.

Finally, significant regional differences in HYV adoption
are observed. Noakhali farmers are more likely to adopt
HYV compared to both the Bogra and Sylhet farmers and the
intensity of adoption of the former is also higher (with the
only exception of the intensity of HYV Boro adoption by Bogra
farms).

Table 3. Output, Variance and Coefficient of
Determination Under Technological
Change in Rice Cultivation: Bogra,
Noakhali and Sylhet Villages of
Bangladesh 1974-75.

Statistic	Season Crop	Aman 1975 Traditional	HYV	Boro 1974-75 Traditional	HYV
Mean output ($\bar{x}$) (maunds per acre)		18.707	26.508	19.742	32.277
Variance (σ^2)		52.777	118.930	104.982	144.948
Coefficient of variation ($\sigma/\bar{x}$)		.388	.411	.520	.373
Number of farms		270	203	176	302

Table 4. Maximum Likelihood Estimation
of Dichotomous Logit Relationship[1]
(Dummy Dependent Variable; 1 if the
farm adopts HYV, 0 otherwise)[2]

Data coverage / Explanatory Variable	Both Seasons		Aman 1975		Boro 1974-75	
	Without Village Dummy	With Village Dummy	Without Village Dummy	With Village Dummy	Without Village Dummy	With Village Dummy
Institutional						
Size of operational holding (acres)	0.7907* (5.1857)	0.9980* (5.0345)	-0.0990§ (-1.5647)	0.1562*** (1.9056)	0.5335* (5.0777)	0.5326* (3.9005)
Percentage of area leased-in	-0.0059*** (-1.8130)	-0.0028 (-0.7561)	-0.0099* (-2.9055)	-0.0027 (-0.6850)	-0.0064** (-2.1174)	-0.0053§ (-1.4125)
Membership in Farmer organisation (dummy)	0.7803* (2.9872)	0.2623 (0.7801)	1.8383* (7.7216)	0.6886* (2.3989)	0.7420* (3.2194)	0.9112* (2.7269)
Demographic						
No. of adult agricultural workers in the household	-0.1982*** (-2.0964)	0.0064 (0.0555)	-0.0010 (-0.0120)	0.0695 (0.6983)	-0.3455* (-3.9389)	-0.1833§ (-1.6648)
Age of the respondent	0.0003 (0.0289)	-0.0105 (-1.0114)	0.0075 (0.9341)	0.0071 (0.7242)	0.0086 (1.0650)	0.0003 (0.0315)
Educational Dummies (Above primary level excluded)						
Illiterate	-0.2703 (-0.7377)	-0.0521 (-0.1250)	-0.6426** (-2.2290)	-0.7503** (-2.1673)	-0.0914 (-0.2953)	0.3578 (0.9518)
Literate	-0.1096 (-0.2331)	-0.0486 (-0.0903)	0.4650 (1.1819)	0.2892 (0.6580)	-0.3880 (-0.9830)	0.0629 (0.1279)
Up to Primary level	0.0733 (0.1601)	-0.1194 (-0.2164)	0.1627 (0.4601)	-0.1932 (-0.4490)	0.0327 (0.0862)	-0.0453 (-0.0958)
Village Dummies (Noakhali excluded)						
Bogra	--	-5.0903* (-4.8587)	--	-4.2929* (-5.7237)	--	-3.5530* (-7.7918)
Sylhet	--	-3.5907* (-3.3542)	--	-5.2489* (-6.8045)	--	-0.9895*** (-1.9700)
Intercept	0.6069 (1.1407)	4.2144* (3.5476)	-0.9719** (-2.0970)	3.0755* (3.4218)	0.3106 (0.6682)	1.6306* (2.3481)
R^2	0.1510	0.3414	0.2457	0.4527	0.1391	0.4205
No. of observations	459	459	456	456	456	456

[1]Description of this statistical technique is presented in the appendix

[2]Figures in parentheses are the t-statistic of the coefficients

*The coefficient is significantly different from zero at the 1% level

**The coefficient is significantly different from zero at the 2.5% level

***The coefficient is significantly different from zero at the 5% level

§The coefficient is significantly different from zero at the 10% level

Table 5. Maximum Likelihood Estimation of
Dichotomous Logit Relationship[1]
(Dummy Dependent Variable = 1 if
the farmer adopts HYV, 0 otherwise;
and Explanatory Binary Variables)
(Standardized value of coefficients)[2]

Explanatory Binary Variables	Data Coverage	Aman 1975		Boro 1974-75	
		Without Village Dummy	With Village Dummy	Without Village Dummy	With Village Dummy
Institutional					
Farm Size Dummies (Small farms, i.e. less than 2.5 acres excluded)					
Large Farms (over 5.5 acres)		-0.0042 (-0.3031)	0.0315* (2.5850)	0.0274* (2.3537)	0.0204*** (1.9140)
Medium Farms (2.5 - 5.5 acres)		-0.0192 (-0.6255)	0.0844* (2.7637)	0.0764* (3.9259)	0.0552* (2.8895)
Tenancy Dummies (Pure tenants excluded)					
Owner farms		0.1884§ (1.6610)	0.0854 (0.6545)	0.0436 (0.6595)	0.1080*** (2.0051)
Owner-cum-tenants		0.0568 (0.7161)	0.0819 (0.9257)	-0.0519 (-1.4073)	0.0395 (1.0441)
Membership in farmer organisations Dummy		0.5517* (7.4870)	0.1795* (2.4229)	0.1358* (3.4750)	0.1505* (3.1756)
Educational Dummies (Above primary level excluded)					
Illiterate		-0.1872** (-2.2921)	-0.2164* (-2.5818)	-0.0241 (-0.5112)	0.0274 (0.5546)
Literate		0.0371 (1.3992)	0.0138 (0.5447)	-0.0125 (-0.8710)	0.0008 (0.0495)
Less than Primary Level		0.0181 (0.6116)	-0.0154 (-0.5089)	-0.0042 (-0.2447)	-0.0042 (-0.2318)
Village Dummies (Noakhali excluded) Bogra		--	-0.6840* (-5.7857)	--	-0.3003* (-7.7432)
Sylhet		--	-1.1446 (-6.9545)	--	-0.0764 (-1.3538)
Intercept		-1.4256* (-3.3234)	3.3462* (3.6073)	0.2043 (0.5294)	0.9251 (1.4667)
R^2		0.2328	.4612	0.1106	0.4101
No. of Observations		456	456	456	456

[1] Description of this statistical technique is presented in the appendix
[2] Figures in parentheses are the t-statistic of the coefficients
* The coefficient is significantly different from zero at the 1% level
** The coefficient is significantly different from zero at the 2.5% level
*** The coefficient is significantly different from zero at the 5% level
§ The coefficient is significantly different from zero at the 10% level

Table 6. Dependent Variable: Percentage Rice Area
under HYV: Bogra, Noakhali and Sylhet
Villages of Bangladesh 1974-75 (Regression Coefficients)[1]

Data coverage / Explanatory variables	Aman 1975 Without village dummy	Aman 1975 With village dummy	Boro 1974-75 Without village dummy	Boro 1974-75 With village dummy
Institutional				
Size of operational holding (acres)	-3.7564* (1.1181)	-0.8027 (0.9116)	-2.1784* (0.8540)	-1.3075 (0.9214)
Percentage of area leased-in	-0.2126* (0.0624)	-0.0084 (0.0474)	-0.0402 (0.0544)	-0.0155 (0.0532)
Membership in farmer organisation (dummy)	26.6083* (4.3623)	5.5055 (3.5582)	13.3257* (3.9268)	4.0465 (4.5577)
No. of adult agricultural workers in the household	-3.3987** (1.4930)	-0.3806 (1.1042)	-	-
Percentage of rice area irrigated	0.1793 (0.1286)	0.3565* (0.0947)	0.3646* (0.0528)	0.3450* (0.0517)
Educational dummies (Above primary level excluded)				
Illiterate	-8.7535*** (4.6974)	-2.5156 (3.9612)	4.9096 (4.8498)	5.9746 (4.6793)
Literate	-	3.9047 (5.0375)	6.0539 (6.5086)	3.7813 (6.2783)
Up to primary level	6.5959 (5.9717)	2.3763 (4.7098)	-3.4545 (5.8119)	-3.3040 (5.6115)
Village dummies (Noakhali excluded)				
Bogra	-	-72.8662* (4.1201)	-	14.4025** (6.3014)
Sylhet	-	-55.3607* (4.7243)	-	-15.4004* (5.2185)
Intercept	45.8626	82.5295	35.7589	47.0853
Mean of dependent variable	38.1096	38.1096	69.4734	69.4734
R^2	0.2124	0.5848	0.1757	0.2412
R^2 (adjusted)	0.1968	0.5729	0.1582	0.2204
F statistic	13.5639	49.1471	10.0466	11.5856
No. of observations	360	360	338	338

[1] Figures in parentheses are the standard errors of the coefficients

* The coefficient is significantly different from zero at the 1% level

** The coefficient is significantly different from zero at the 2.5% level

*** The coefficient is significantly different from zero at the 5% level

The coefficient is significantly different from zero at the 10% level

Table 7. Farm Size and HYV Adoption: Bangladesh 1974-75

Season	Both Seasons		Aman 1975				Boro 1974-75			
Farm-Size category (acres)	Total No. of farms	Adopters (% of farms)	Total No. of farms	Adopters (% of farms)	Intensity of adoption (% rice area under HYV)	Index of Participation pation [h]	Total No. of farms	Adopters (% of farms)	Intensity of adoption (% rice area under HYV)	Index of Participation pation
0.0 - 0.5	91	59.3	91	42.9	40.16	17.23	89	49.4	44.12	21.80
0.5 - 1.0	96	68.8	96	43.8	30.44	13.33	96	61.5	52.11	32.05
1.0 - 1.5	72	72.2	72	41.7	27.92	11.67	71	56.3	43.81	24.67
1.5 - 2.0	61	83.6	61	42.6	27.84	11.86	61	70.5	57.80	40.75
2.0 - 2.5	30	86.7	30	46.7	26.33	12.29	30	73.3	49.00	35.92
2.5 - 3.0	25	88.0	25	44.0	25.71	11.31	25	88.0	70.54	62.08
3.0 - 3.5	31	90.3	31	48.4	28.33	13.71	31	77.4	49.37	38.21
3.5 - 5.5	31	100.0	31	41.9	19.76	8.30	31	87.1	57.47	50.06
5.5 and above	22	100.0	22	54.5	23.18	12.63	22	90.9	56.35	51.22
All farms	459	76.7	459	44.0	29.89	13.15	456	66.0	51.16	34.06
Chi Square (x^2) or F-statistic	—	43.93	—	1.63	1.16	—	—	35.41	1.523	—
Level of Significance of x^2 or F-statistic	—	.0001	—	.99	.322	—	—	.0001	.147	—
Pearson's Correlation Coefficient (R)	—	.302	—	.035	—	—	—	.262	—	—
Level of Significance of R	—	.0001	—	.2285	—	—	—	.0001	—	—

[h]Index of participation is derived by simply multiplying the crude adoption rate and intensity of adoption (i.e. multiplying one ratio by the other).

[a]Actually refers to adoption in either season.

Table 8. Tenancy and HYV Adoption: Bangladesh 1974-75

Season	Both Seasons		Aman 1975				Boro 1974-75			
	Total No. of farms	Adopters (% of farms)	Total No. of farms	Adopters (% of farms)	Intensity of adoption (% rice area under HYV)	Index of Participation	Total No. of farms	Adopters (% of farms)	Intensity of adoption (% rice area under HYV)	Index of Participation
Owner-cultivator	240	79.6	240	50.0	36.73	18.37	238	73.1	56.87	41.57
Owner-cum-Tenant	16	76.4	161	41.6	22.15	9.27	161	58.4	43.04	25.14
Tenant	48	64.6	48	22.9	19.57	4.48	47	59.6	52.17	31.09
All Categories	449	76.8	449	44.1	29.67	13.08	446	66.4	51.41	34.14
Chi-Square (x^2) or F-statistic	-	5.05	-	12.53	8.23	-	-	10.42	5.031	-
Level of Significance	-	.08	-	.0019	.0003	-	-	.0055	.0069	-

[a] Index of participation is derived by simply multiplying the crude adoption rate and intensity of adoption (i.e. multiplying one ratio by the other).

Table 9. Participation in Farmers Organisations and HYV Adoption: Bangladesh 1974-75

Season Participation Status	Both Seasons		Aman 1975			Boro 1974-75		
	Total No. of farms	Adopters (% of farms)	Adopters (% of farms)	Intensity of Adoption (% rice area under HYV)	Index of Participation [a]	Adopters (% of farms)	Intensity of Adoption(% rice area under HYV)	Index of Participation [a]
Member	248	59.9	64.5	44.85	28.93	74.6	58.84	43.89
Non-member	212	40.1	19.8	12.31	2.44	55.5	42.13	23.38
Both categories	460	76.5	43.9	29.89	13.12	65.9	53.16	35.03
Chi-Square (x^2) or F statistic	—	20.92	90.94	87.04	—	17.55	17.59	—
Level of Significance	—	.0001	.0001	.0001	—	.0001	.0001	—

[a] Index of participation is derived by simply multiplying the crude adoption rate and intensity of adoption (i.e. multiplying one ratio by the other).

Table 10. HYV Adoption and Farm Size
over time (1968-1975)
Bangladesh
(percentage of cultivators
under each farm size)

Farm Size (acres)	Total No. of farms	1968	1969	1970	1971	1972	1973	1974	1975
0.0 - 0.5	91	16.5	20.9	27.5	41.8	44.0	48.4	55.0	58.3
0.5 - 1.0	96	14.6	25.0	33.3	42.7	49.0	55.3	61.6	66.8
1.0 - 1.5	72	19.4	29.1	33.3	43.0	48.6	58.3	66.6	72.2
1.5 - 2.0	61	18.0	22.9	47.5	57.3	68.8	77.0	81.9	90.1
2.0 - 2.5	30	20.0	43.3	56.6	69.9	73.2	83.2	86.5	89.8
2.5 - 3.0	25	28.0	32.0	32.0	48.0	56.0	68.0	76.0	84.0
3.0 - 3.5	31	12.9	29.0	51.6	71.0	83.9	83.9	87.1	90.3
3.5 - 5.5	31	25.8	35.5	42.0	61.4	77.5	90.4	93.6	96.8
5.5 + above	22	13.6	27.2	49.9	68.1	86.3	90.8	95.3	99.8
All categories	459	17.9	27.3	36.9	49.8	57.4	64.6	70.5	75.5

Table 11. Tenancy and HYV Adoption over time (1968-1975) : Bangladesh
(percentage of cultivators under each category)

Farm Category	Total No. of farms	1968	1969	1970	1971	1972	1973	1974	1975
Owner-Cultivator	240	22.1	33.4	41.7	54.6	62.5	70.4	75.0	78.8
Owner-cum-tenant	161	16.1	23.6	34.8	49.7	56.5	63.3	70.1	75.1
Tenant	48	6.3	14.6	15.0	21.3	29.6	33.8	42.1	52.5
All categories	449	18.3	27.9	37.5	50.4	58.0	65.1	70.9	75.8

Table 12. Participation in Farmers Organisation
and HYV Adoption over time (1968-1975):
Bangladesh
(percentage of cultivators
under each category)

Participation Status	Total No. of Farms	1968	1969	1970	1971	1972	1973	1974	1975
Member	257	23.3	35.8	47.1	59.6	65.0	69.4	67.2	69.9
Nonmember	368	6.3	9.3	13.4	20.7	26.4	31.6	33.8	38.1
All categories	625	13.3	20.2	27.0	36.4	42.0	47.3	51.8	55.5

Chapter 3

Agrarian Structure, Production Efficiency and
Technological Change

Introduction

The relationship between the prevailing agrarian
structure and farm productivity is studied in this chapter.
The influence of the two most important institutional factors,
farm size distribution and tenancy, on land productivity is
closely examined. Moreover, for an overall assessment of
social efficiency in a labour-abundant country like Bangladesh,
where land is the most scarce factor, the share of farm land
double - or multiple-cropped would also have to be considered.
The relationship between the agrarian structure and farm
productivity is also reviewed under technological change.
Essentially, this will enable us to examine the implications
for agrarian reform policies when dynamic factors (such as
technological change) are taken into account. Before we proceed
with the empirical verification of these relationships, the
analytical and theoretical framework are briefly discussed in
the following three sections.

Returns to Scale and Technological Change

Before embarking on an assessment of the relationship
between farm size and land productivity, it is necessary to
ascertain the nature of the returns to scale. The issue of
"technical input-output" efficiency has to be distinguished
from the broader question of efficiency of resource utilization. The first refers strictly to the engineering relationship

of production per unit of inputs actually used in the production
process. The second considers as well the selection of those inputs, and particularly the degree of utilization of the available
land resource (considered below) and the related use of labour
(considered in the next chapter). Thus, for crop production
only, the land actually cultivated enters the "technical
production function" in an evaluation of returns to scale.
But, for over-all assessment of social efficiency, the share
of land double or multiple cropped must also be considered.

For determining the returns to scale in Bangladesh rice
cultivation, a Cobb-Douglas production function of the following
form is fitted.

$$\log Q = \log A + \alpha \log X_1 + \beta \log X_2 + U$$

where Q is the rice output in maunds, A is the constant term,
X_1 is the labour input in man-days (standard 8-hours), X_2 is
the area sown under the crop in acres and U is the error term. The elasticity
estimates presented in Table 13 reveal that the input elasticity
for labour is nearly the same (between .28 to .30) for both
the traditional and HYV rice during Aman. However, during
the Boro season, the input elasticity for labour is .21 for
HYV rice as compared to .50 for the traditional variety. The
striking point is the much larger input elasticity for land
following technological change (.73 for HYV Aman as compared
to .59 for traditional Aman; and .76 for HYV Boro as compared
to .54 for traditional Boro).[1]

Test of the hypothesis of constant returns to scale
was undertaken for both the traditional and HYV rice crops
covering both the seasons by means of t-test[2]

$$H_o: \quad \alpha + \beta = 1$$

---

[1] An interesting similarity of the land input elasticity
is observed for the green revolution area of the Pakistan
Punjab [11, p.67]. This represents land augmenting technological change.

[2] The t statistic was computed as

$$\frac{\alpha + \beta - 1}{\sqrt{s_\alpha^2 + s_\beta^2 + 2 \text{ Est.Cov}}} \sim t_{n-2}$$

An important question to look at is whether the introduction of the seed-fertilizer (Green Revolution) technology
alters the returns to scale in agriculture. The above
hypotheses is not rejected for any of the crops. It is
systematically found that returns to scale are not significantly
different from constant returns. In other words, it can be
argued that as there are no economies of scale in actual
farming operations, there is not likely to be a sacrifice
of potential efficiency from land redistribution. This
argument is as much valid under dynamic agriculture as it
is under a traditional agricultural setting.

Agrarian Structure, Productivity and Technical Change

The important question being investigated here is whether
the prevailing agrarian structure with unequal land distribution
would lead to under-utilization of land and therefore low land
productivity in larger farms. It has been argued by some that
land scarce countries like Bangladesh could expect little
production gain from land redistribution, because the division
of "large" farms, already of a modest size by international
standards (due to land scarcity), into the very small parcels
that would be necessary to absorb the rural population may cause
stagnation rather than raise output. We have already seen
that for all crops, traditional and new varieties, returns to
scale is not significantly different from constant returns.
Therefore, there is no reason for planners to fear losses of
production efficiency from the division of existing large
farms into small farms even under conditions of technological
change.

The implication of the existence of constant returns to
scale is that output per acre should be constant across farm
sizes, that is, it should be independent of the size of the

farm.   However, in a wide range of countries, an inverse
relationship between land productivity and farm size has
been observed in conjunction with generally constant returns
to scale in agriculture.[1] Our interest here is to examine
the extent to which this phenomenon is also observed in
Bangladesh agriculture.

In investigating the relationship between farm size
and land productivity an appropriate functional form has to
be chosen. As seen from the above, theory dictates no
(negative) relationship and therefore no estimating form.
However, an estimating form which best describes the data
has to be chosen. The basic forms that have been tried in
the literature are three: the linear, log-log, and semi-
log. The form used by us are the semi-log and the log-
log (models 1 and 2); that is

$$(1) \quad y = a + b \log A + U$$
$$\text{and} \quad (2) \quad \log y = \log a + b \log A + U$$

where y is rice output per acre (of land planted to rice) in maunds,
A is size of the operational holding in acres and U is the error term.

Based on data available for various regions of India,
reservations have been expressed about the universal validity
of the inverse relationship between land productivity and
farm size [20]. Another note of caution has been delivered
on the effect of aggregation or pooling of data pertaining to,
different villages. In such a situation there is a risk of
falsifying the dependence of the productivity on farm size
in both directions [20]. The statistical error on account of

---

[1]For an excellent review of evidence see [13 and 20].

the pooling of data from different villages could result
in either a spurious inverse relationship or the intra-
village inverse relationship being obliterated.[1]

The second important institutional factor being considered
is share cropping. Land tenancy may depress productivity and
labour use on account of the tenurial insecurity as well as
the prevalent crop-sharing system. The traditional viewpoint
is that share-cropping causes inefficient allocation of
resources because, unless landowner and sharecropper share all
input costs in the same proportion as output, the sharecropper
will apply below optimal levels of variable inputs since he
will equate the full marginal cost of such inputs with his
share of their marginal product, not with the total marginal
product [See 13]. This view has been followed by a spree of
controversy at the theoretical level. For example Cheung [22]
argued that the nature of tenancy would have no effect on
output, provided the landlord was in a position to dictate the
intensity of variable inputs. This neoclassical contention
has not gone unchallenged [12]. However, given that the share-
cropping (50% of output received by landlord) is by and large
the only tenancy system that prevails in Bangladesh agriculture,
some negative effect of tenancy on output should be observed.

---

[1]For example, if within a single village, productivity
is invariable with respect to farm size but the productivity
is different in different villages, and if the villages with
higher productivity have small average farm sizes, then putting
together data from different villages would reveal a spurious
inverse relationship. If, on the other hand, average
productivity is the same in every village and if within each
village the inverse relationship holds, then if the different
villages have different average ranges of farm sizes, the
intra-village inverse relationship would become obliterated
in the process of pooling of inter-village data.

From the policy point of view, the relevant question is how severely the existence of sharecropping lessens the productive potential of agriculture (if at all), and, accordingly, how large the output gains might be expected of a shift from sharecropping to farm ownership (land to the tillers).

It is of major policy interest to consider whether dynamic factors such as the introduction of the green revolution technology alters the relationship between land productivity and the agrarian structure. No empirical evidence appears to be available to shed light on this important question. The few observations that have been made are speculative and conjectural rather than being based on any rigorous and systematic treatment of data.[1]

Agrarian Structure and Cropping Intensity

Finally, land is not a timeless, static concept. The 12.3 million acres of cultivated land that Bangladesh possesses is as good as 36.9 million acres, for instance, if it is used three times a year. That is to say, it is the degree of utilisation which determines the effective amount of land resource of a country. Therefore, if the intensity of cropping is negatively related to farm size and with tenancy, in a sense it could be concluded that the prevailing agrarian structure is holding back the growth of Bangladesh's land endowment. On the other hand, it would be of considerable interest to investigate if organising farmers into irrigation groups around a pump leads to higher cropping intensity. On an average the index of cropping intensity in Bangladesh at present is about 148.5. This means that over one-half of the net sown area produces only one crop in a year.

---

[1]A beginning has however been made by Berry and Cline [13].

Empirical Verification

Having briefly reviewed the theoretical and analytical framework we undertake the empirical verification of the relationship between the agrarian structure and productivity with the help of multiple regression analysis of farm-level cross-sectional variation in productivity per acre for both traditional and HYV rice varieties covering the Aman and Boro seasons using both pooled (aggregate) and intra-village (disaggregate) data.

Agrarian Structure and Productivity

Average rice output during the Boro season was 32.3 maunds per acres for HYV rice as compared to 20 maunds for the traditional variety. During the Aman season the rice output per acre for HYV was 26.5 maunds as compared to 18.7 maunds for the traditional variety. These average figures show the wide differences in land productivity with variations in crops and seasons. Tables 2-5 present the results of the multiple regression analysis of the determinants of productivity per acre for both seasons and crop varieties.

Traditional rice output per acre during the Boro season is <u>negatively</u> related to farm size and it is found valid for both pooled and disaggregate data (Table 15). With the introduction of the green revolution technology during Boro, this relationship appears to have remained <u>unaltered</u> with respect to only one village and is no longer significant in terms of pooled data (Table 14). However, this relationship is not universal for both crops. Land tenancy has a significant <u>depressing</u> effect on productivity of Boro traditional rice crop in respect of only one village; with the introduction of the green revolution technology, the negative relationship between productivity and tenancy appears to have <u>strengthened</u> although not more universal than before. This is well established in terms of both pooled and disaggregate data.

Farms differ in land quality and this could have an effect on potential output of a farm. An explanation of interfarm

productivity differences in terms of land quality has to be
supported by the observation that land quality worsens with
farm size and/or land tenancy and these differences account for
all the variations in output per acre. In the following discussion, the various arguments for quality differences are
considered and incorporated into the framework of the basic
model (equations 1 and 2). Since inherent differences in land
quality do exist, they have to be taken into account in any
explanation of farm productivity differences. First, price of
land is used as proxy for land quality in the sense that it
measures soil fertility. However, this accounts for only
exogeneous differences in soil fertility. Therefore, endogeneous
determinants of land quality have also to be considered. One
variation on the theme of land quality is that endogeneous
determinants such as man-made improvements in the form of
irrigation and fertiliser application can affect the fertility
of the soil. Another important question to consider is whether
these inverse relationships are valid when tractor farms are
distinguished from bullock farms. However, it is to be
recognised that exogeneous and endogeneous determinants of
soil fertility are difficult to separate.

Therefore inter-farm exogeneous differences in soil fertility
and endogeneous differences such as in irrigation, fertiliser
application and mechanisation level in preparatory tillage
are introduced in the basic model as follows.

(3) $y = a + b \log A + cT + dP + eI + fN + gM + hR$

and (4) $\log y = a + b \log A + cT + dP + eI + fN + gM + hR$

where, as before, y and A respectively denote rice output per
acre and size of operational holding in acres; T denotes
percentage of farm area leased-in, P denotes price of an acre
of rice land in Bangladesh Takas, I denotes percentage of farm
area irrigated, N denotes chemical fertiliser in seers applied
per acre, M denotes farmyard manure in maunds applied per acre
and R denotes the mechanisation level in preparatory tillage
(dummy). As can be seen, price of land, P is introduced into
the basic model in an additive manner, thereby implying that unit
increase in price/acre results in constant addition to output/acre.
Increases in price will always result in increase in output/acre.

Therefore, the model makes the important assumption that P
captures differences in land input quality, so that the same
amount of physical land input results in higher output when land
price is higher.

The above model now being more comprehensive essentially
answers the question: given exogeneous differences in soil fertility and endogeneous differences in irrigation, fertiliser-use and
mechanisation level, does one still observe significantly negative
coefficients for b and c. As can be seen from Tables 14 and 15
after controlling for soil-fertility and for these man-made land
improvement factors (irrigation and fertilizer application) and
mechanisation level, the above conclusions about the
relationship between farm size and productivity and that between
land tenancy and productivity remain unaltered. Land productivity
has a negative relationship with the exogeneous soil fertility factor
(land price) for traditional Boro crop (significant only with
respect to pooled data); it is also negatively related to P following
the use of the Green Revolution Technology (it is significant for
pooled data and for the Bogra village data). As expected, land
productivity is positively related to irrigation, fertiliser
application and farm mechanisation (in one village) in respect
of Boro HYV crop (Table 14). None of these factors have a significant relationship with productivity under the traditional Boro
crop (Table 15).

During the Boro season significant regional differences in land
productivity are observed for both the traditional and HYV crops.
Productivity in the Bogra and Sylhet villages appear to be significantly higher than that of the Noakhali village.

During the Aman season an inverse relationship between productivity
and farm size is observed only for the traditional rice crop (Tables
16 and 17) and this is valid both for pooled and disaggregate data,
although it is not completely universal. Land tenancy appears to
have a depressing effect in productivity in respect of the traditional crop and this is observed for both pooled and disaggregate
data although it is not fully universal; no such effect is observed
for the HYV. Land productivity is negatively related to hired proportion of total labour used per acre for the traditional crop in one
village and positively related in another. As expected, land productivity has a significant positive relationship with the price of

land for the traditional Aman crop and this is valid for both pooled
and disaggregate data although it is not universal. Use of land
improvement factors (irrigation and fertiliser) is positively related
to productivity of the traditional crop. As before, controlling
for exogeneous (soil fertility) and endogeneous man-made land
improvement factors and mechanisation level, the above conclusions
about the relationships between productivity and farm size and
tenancy remains <u>unaltered</u>.

Regional difference in land productivity during Aman
is not so common with only the Sylhet village having significantly
higher productivity over that of Noakhali.

Intensity of Input Use

Having established that productivity is inversely
related to farm size and tenancy (though not fully universal)
it is of interest to investigate if this is the result of
small farmers' or owner operators' using more inputs per acre.
The intensity of factor use and its variability with farm size
and tenancy is explored in this section.

In rice cultivation, the first important input after
land is labour. In the next chapter (on labour utilisation
under technical change) it will be seen that small farms put
in more labour input per acre. For the smaller farms, dependent
largely on family labour, the imputed cost of labour is less
than that for the larger farms hiring labour at the market
wage rate. Another reason why labour intensity is higher on
family-based smaller farms is better quality of management that
can be applied; while larger farms have to incur supervision costs
for hired labour. Moreover, the need for small farmer's
survival drives him to more intensive effort. A poor peasant
family, depending on a small piece of land, would not only
ignore any marginal productivity calculations in so far as
family labour is concerned, he would employ hired labour
whenever necessary to supplement the family labour and in so
doing would pay no heed to marginal productivities. He would
try to improve the quality of the land by small-scale irrigation
and other such means as can be procured with the help of

labour. Land tenancy appears to depress labour-use (see next chapter) on account of the tenurial insecurity and the prevailing crop-sharing system.

The second important input whose use intensity we examine is chemical fertilizer. Fertilizer dose applied on an acre of HYV crop varies between 43 to 45 seers for nitrogen, 6.5 to 7.5 seers for potash and 28 to 31 seers for phosphate. Nitrogen fertiliser dose applied is <u>negatively</u> associated with farm size and tenancy and <u>positively</u> with membership in farmers' organisations (Table 18). Similar relationships with the agrarian structure are observed in the per acre application of the phosphate fertiliser. Application of potash fertiliser appears to be positively related to farm size and does not have any significant relationship with either tenancy or participation in farmer organisation. Participation in farmer organisation offers easier access to cheaper institutional credit and fertiliser and irrigation inputs at subsidised rates. After controlling for participation in farmers' organisations, the above conclusions remain <u>unaltered</u>.

Another important factor to consider is weeding intensity. On an average the frequency of weeding a HYV plot is twice that of a traditional rice plot during both seasons. On the whole the frequency of weeding is <u>negatively</u> related to farm size and tenancy and <u>positively</u> related to membership in farmers' organisation (Table 19). The only exception being the HYV Aman crop in which statistically no significant relationship is observed.

These results provide a strong explanation for the inverse relationships with farm size and with tenancy. A considerable portion of the inter-farm differences in output per acre by farm size and by tenancy is due to the difference in the intensity of input use.

Agrarian Structure and Cropping Intensity

So far we tried to determine empirically the relationship between the agrarian structure and productivity of land which is actually put into cultivation. In this section we make an assessment of the prevailing agrarian structure on the share

of available land double- or multiple-cropped (that is,
cropping intensity or multiple cropping index) and possible
explanations for the observed relationships are offered.

As before the multiple regression analysis of farm
level cross-sectional data shows that cropping intensity is
negatively associated with farm-size both for adopters and non-adopters
of HYV (Table 20). Tenancy does not appear to have any significant effect on
cropping intensity. Factors such as farmers' participation in irrigation
groups and in addition, the percentage of farm area irrigated
are usually important determinants of cropping intensity. An
explanation of interfarm differences in cropping intensity has
to be supported by the observation that both membership in
irrigation societies and percentage of farm area irrigated
decreases with farm size and these differences account for
all the variations in cropping intensity. The important
question is that: given differences in the use and availability
of irrigation water does one still observe a significantly
negative relationship between farm size and cropping intensity.
As can be seen from Table 20, after controlling for the use of
irrigation water, our conclusions about the significant inverse
relationship between farm size and cropping intensity remains
unaltered. Cropping intensity appears to be positively
associated with membership in irrigation groups and proportion
of are irrigated .

Even after controlling for regional differences in
cropping intensity, the inverse relationship between farm size
and cropping intensity remains unaltered; although the positive
relationship between irrigation and cropping intensity is
no longer observed. Significant regional differences in
cropping intensity exists; while cropping intensity in the Bogra
village is significantly higher than that of the Noakhali
village, that in Sylhet is significantly lower.

Several explanations have been put forward for the inverse
relationship between cropping intensity and farm size.[1] A common

---

[1]The multiple cropping issue is more a reflection of the general
land-use intensity and farm size relationship.

explanation is that the farmer is not interested in income
beyond a certain level, so he leaves the land idle after that
level of income is reached. A rectangular hyperbola situation
is assumed under which cropping intensity declines as farm
size goes up keeping the total income constant. This explanation of big farmers being content when a certain level of
income from agriculture is achieved has been refuted on the
ground that they try to increase their total income through
various nonagricultural activities.

Another explanation put forward is that the big farmer
(supplying a sizeable part of the local market) exercises
his monopoly power to restrict output. This view has also
been challenged on the ground that a 'monopolist' farmer
may limit the production of rice in the hope of a better
price; but why should he not follow up the rice crop with,
say, potato.

A third reasoning put forward is that the big farmer may
not find multiple cropping profitable. This point is explained

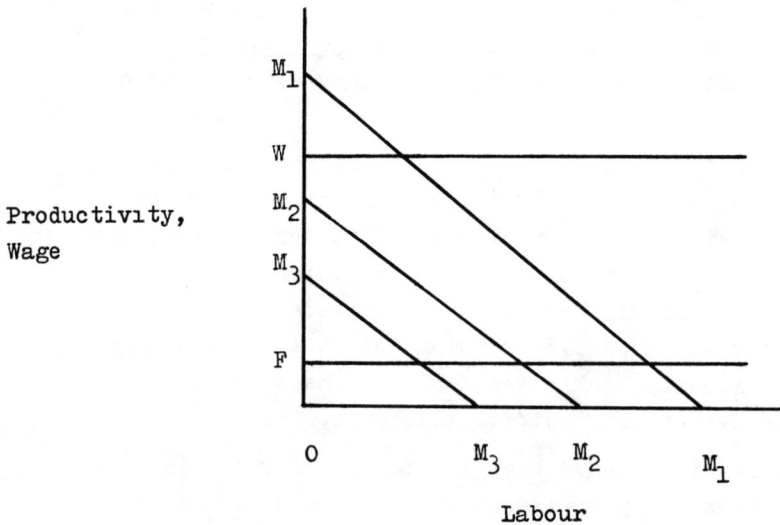

by means of the above figure. Let M_1M_1, M_2M_2 and M_3M_3 be
the curves of marginal productivity of labour for three
consecutive crops, respectively. The big farmer employs
hired labour at wage rate OW. Clearly, the **second** and **third**
crops are not profitable for him to produce. A small farmer,
on the other hand, cultivates land solely or mainly with his
own family labour whose wage rate is practically zero, or
say OF in the figure. In this case the small farmer would
raise three crops on his land, while the big farmer would
leave his land idle after the first crop.

A fourth explanation put forward contends that the
big farmer has a multiplicity of channels of profit-making
[58]. He cultivates his land with hired labour, lends money
to small farmers and agricultural labourers at usurious rates
of interest, engages in trading in foodgrains and other agricultural
products, and brings industrial goods from the town to the
village market. All these four modes of extracting profit
are exercised by him throughout the year in a certain 'optimal'
sequence; and cultivation of land appears on his agenda only
for a brief period; at other times of the year his land remains
fallow.

It is possible that the big farmers in the survey area
may not find multiple cropping profitable in view of their
high reliance on the use of hired labour (a significant
positive association exists between farm size and hired
proportion of total labour used as can be seen in the following
chapter). Moreover, the hypothesis of multiplicity of channels
of profits making available to large farmers is likely to be
valid in the survey villages. A significant (at 1% level)
positive relationship between farm size and total number of
occupations is observed.[1] The big farmers are also found to

---

[1]The linear regression equation fitted yielded the following
result: Number of occupations= 1.64 + .068 Farm Size, R^2 = .03
N = 459 observations and the figure in parenthesis is the
(.019)
standard error of the regression coefficient.

be actively engaged in the trading of food grains as is confirmed by
a significant positive relationship between farm size and the
proportion of rice output sold in the market during both the Aman
and Boro seasons (see Chapter 5).

Summary and Conclusions

By fitting a Cobb-Douglas production function it is systematically
found that the returns to scale in Bangladesh agriculture are not
significantly different from constant returns and this finding is
equally valid for traditional and the high yielding variety rice crops.
The implication of the existance of constant returns to scale is that
output per acre should be constant across farm sizes, that is independent of the size of the farm.   However, under a traditional agricultural
setting, an _inverse_ relationship between land productivity and farm
size has been observed in conjunction with constant returns to scale
in Bangladesh agriculture.   The introduction of the green revolution
technology has either _weakened_ (during Boro) or _eliminated_ (during
Aman) this inverse relationship.

The multiple regression analysis of the interfarm variations in
productivity shows that rice output per acre is _negatively_ associated
with farm size and tenancy.   With the introduction of the green
revolution technology the inverse relationships between productivity
and farm size weakened during Boro and disappeared during Aman;   and
the inverse relationship between productivity and tenancy is still
valid for Boro season;   during the Aman season this relationship does
no longer appear to be significant.

Productivity is _negatively_ related to the exogeneous land quality
factor (soil fertility measured by the price of land) during the Boro
season (in terms of pooled data and in one village with respect to HYV
Boro).   This relationship is found to be _positive_ in the case of
traditional Aman crop (in terms of pooled and disaggregate data).
Endogeneous man-made land improvement factors such as irrigation and
fertiliser application which affect land quality have a significant
positive impact on land productivity.   The inverse relationship has been
established both in terms of pooled (aggregate) and intravillage (disaggregate) data and as such the conclusions are free of the statistical

fallacy that one runs into by relying on results obtained only from
aggregate or pooled data.   For none of the crops (HYV and traditional)
the inverse relationship is fully universal.

After controlling for exogeneous land quality factors like soil
fertility (measured by the price of land, P) and endogeneous land
improvement factors and the level of mechanisation in land tillage,
the above conclusions about the relationships between land productivity
and farm size and between productivity and tenancy remain unaltered.
Influence of the mode of labour use on land productivity is not clear.
While in most crops no relationship was observed, land productivity in
the case of the traditional Aman crop was positively related to the
hired proportion of total labour use in one village, it was negatively
related in another.

Significant regional differences in productivity are observed for
both crops during the Boro season and for the HYV crop during the Aman
season.

In sum it can be concluded that although the inverse relationship
between productivity and farm size and between productivity and tenancy
does not appear to be a fully universal law, this feature occurs repeatedly in different combinations of crop varieties and seasons in different
parts of the country.[1]   Furthermore, these relationships are established
after controlling for land quality differences whether in endogeneous
(irrigation and fertilisation) or exongeneous (soil fertility) form.
This is largely the result of higher intensity of individual factor use
(like labour, fertiliser, etc.) by small farmers.   Therefore, it can
be concluded that considerations of dynamic factors such as technological
change, need not reverse the policy implications that arise from the
static output gains to be achieved from a policy of land redistribution
and other programmes favouring the small farm sector.   It can be

---

[1] An inverse relationship between farm size and productivity was
also observed for Dinajpur and Mymensingh villages of Bangladesh [39].
On the contrary, in some villages surveyed in Sind and Punjab provinces
of Pakistan, a positive relationship was observed [45].   The evidence
from India suggests a mixed scenario like the overall conclusions
relating to farm size and productivity being drawn for this study [20].
A survey of farms from Mymensingh, Rangpur and Dinajpur districts of
Bangladesh found share-cropping to have a depressing effect on
productivity [41] whereas another survey of Mymensingh and Dinajpur
villages of Bangladesh found no significant relationship [39].

concluded that share-cropping lessens the productive potential of Bangladesh agriculture and output gains may be expected as a result of a shift from share-cropping to farm ownership.

Finally, a significant <u>negative</u> relationship is observed between cropping intensity and farm size[1] both for adopters and nonadopters. No significant relationship is observed to exist between <u>cropping intensity and tenancy</u>. <u>The inverse</u> relationship between cropping intensity and farm size is most convincingly established when it is seen that even after controlling for the availability (participation in irrigation pump groups) and use (percentage area irrigated) of irrigation water, this <u>inverse</u> relationship remains unaltered. One important reason for the existence of this negative relationship is that the big farmers in the survey area were found to be engaged in a multiplicity of channels of profit making and the cultivation of land appears on their agenda only for a brief period; at other times of the year their land remains fallow.

Tenancy prevents the farms from being integrated into the monetised market economy and tends to leave them in the subsistance sector.

---

[1] Such a relationship is observed for villages surveyed in the Mymensingh and Dinajpur districts of Bangladesh [39]; the existance of this inverse relationship is also observed in Indian agriculture [58].

Table 13. Estimates of the Input Elasticities of the Cobb-Douglas Production Function: Bangladesh 1974-75a,b

Crop	Aman 1975				Boro 1974-75			
	HYV Rice		Traditional Variety		HYV Rice		Traditional Variety	
Coefficient	Un-restricted	Restricted	Un-restricted	Restricted	Un-restricted	Restricted	Un-restricted	Restricted
Labour α	.2766* (.0839)	.2714* (.0800)	.2958* (.0645)	.2966* (.0670)	.2130* (.1553)	.2288* (.0538)	.4957* (.0849)	.5164* (.0820)
Land β	.7297* (.0804)	.7286	.5909* (.0689)	.7034	.7576* (.0553)	.7712	.5408* (.1016)	.4836
$\alpha + \beta$	1.006	1.000	.887	1.000	.971	1.000	1.037	1.000
Returns to scale (t-test) for H_o: $\alpha + \beta = 1$ c	Constant (t=.0515)		Constant (t=1.203)		Constant (t=.3713)		Constant (t=.2797)	
Intercept	.888	.896	.716	.730	1.050	1.024	.3485	.3052
R^2	.857	.055	.838	.069	.847	.058	.847	.194
R^2(adjusted)	.856	.050	.837	.066	.846	.055	.845	.189
F statistic	591.29	11.51	677.85	19.59	818.43	18.12	453.19	39.63
No. of observations	200	200	265	265	298	298	166	166

*The coefficient is significantly different from zero at the 1% level

a Figures in parentheses are the standard errors of the coefficients

b The restriction assigned is $\alpha + \beta = 1$; the regression equation was estimated only in terms of α after substituting $(1-\alpha)$ for β and as a result the significance test is observed only for the labour coefficient (α) in the restricted case.

c The t-statistic is computed as follows:

$$\frac{\alpha + \beta - 1}{\sqrt{s_\alpha^2 + s_\beta^2 + 2 \text{ Est. Cov }(\alpha, \beta)}}$$

Table 14. Dependent Variable: HYV Rice Output (maunds)
per acre: Bangladesh 1974-75 Boro Season
(Regression coefficients)[1]

Data Coverage / Explanatory Variables	Three Villages Aggregate Data		Bogra Village	Sylhet Village	Noakhali Village
	Without Village Dummy	With Village Dummy			
Institutional					
Log of the size of operational holding	-0.3535 (1.7098)	-2.1963 (1.8168)	-2.0742 (6.5210)	0.4878 (2.3304)	-6.7621* (2.7135)
Percentage of area leased-in	-0.0446*** (0.0208)	-0.0478** (0.0203)	-	-0.0344 (0.0237)	-0.0678§ (0.0362)
Hired proportion of labour-days applied	0.0170) (0.0241)	0.0166 (0.0236)	-	0.0358 (0.0298)	-
Land Quality Determinants					
Exogenous					
Price of land (Takas per acre)	-0.0003* (0.0001)	-0.0001 (0.0001)	-0.0020* (0.0006)	0.0001 (0.0002)	0.0000 (0.0001)
Endogenous					
Percentage of area irrigated	0.0314 (0.0215)	0.0139 (0.0213)	0.0803 (0.1530)	0.0062 (0.0241)	0.0136 (0.0354)
Nitrogen fertilizer dose applied (seers per acre)	0.1450* (0.0251)	0.1347* (0.0258)	0.2517* (0.0576)	0.0139 (0.0412)	0.1591* (0.0398)
Farmyard manure applied (maunds per acre)	-0.0059 (0.0091)	-0.0194*** (0.0094)	-0.0140 (0.0133)	-0.0333§ (0.0188)	-0.0353 (0.0512)
Mechanisation level in preparatory tillage (dummy)	-	2.0998 (2.1900)	-	-8.2106 (5.9455)	4.1762*** (2.1119)
Village Dummies (Noakhali excluded)					
Bogra	-	12.3208* (3.2662)	-	-	-
Sylhet	-	5.1504 3.7026	-	-	-
Intercept	29.8230	24.6202	63.0234	29.1832	17.8525
Mean of dependent variable	32.2771	32.2771	41.1054	31.1562	30.5496
R^2	0.1308	0.1842	0.4728	0.0676	0.2410
R^2 (adjusted)	0.1101	0.1561	0.3952	0.0195	0.1819
F statistic	6.3189	6.5690	6.0975	1.4050	4.0815
No. of observations	302	302	40	164	98

[1] Figures in parentheses are the standard errors of the coefficients
* The coefficient is significantly different from zero at the 1% level
** The coefficient is significantly different from zero at the 2.5% level
*** The coefficient is significantly different from zero at the 5% level
§ The coefficient is significantly different from zero at the 10% level

Table 15. Dependent Variable: Log of Local Variety Rice
Output (maunds) per acre: Bangladesh 1974-75
Boro season

Data Coverage / Explanatory Variables	Three Villages Aggregate Data		Noakhali Village
	Without Village Dummy	With Village Dummy	
Institutional			
Log of the size of operational holding	-0.0143 (0.0449)	-0.0747 § (0.0454)	-0.1599§ (0.0939)
Percentage of area leased-in	0.0005 (0.0006)	0.0003 (0.0005)	-0.0038*** (0.0018)
Hired proportion of labour-days applied	0.0006 (0.0006)	0.0008 (0.0006)	-
Land Quality Determinants			
Exogenous Price of land (Takas per acre)	-0.00001* (0.00000)	-0.00000 (0.00000)	-0.00001 (0.00001)
Endogenous Percentage of area irrigated	0.0006 (0.0005)	0.0002 (0.0005)	0.0009 (0.0012)
Nitrogen fertilizer dose applied (seers per acre)	-0.0016 (0.0011)	-0.0007 (0.0011)	-
Farmyard manure applied (maunds per acre)	-0.0003 (0.0003)	-0.0007 (0.0005)	0.0026 (0.0024)
Mechanisation level in preparatory tillage (dummy)	-0.0197 (0.0535)	0.0553 (0.0543)	0.0860 (0.0697)
Village Dummies (Noakhali excluded)			
Bogra	-	0.3934** (0.1740)	-
Sylhet	-	0.2939* (0.0763)	-
Intercept	1.3467	1.0375	1.1586
Mean of dependent variable	1.2666	1.2666	1.0542
R^2	0.2845	0.3569	0.2437
R^2 (adjusted)	0.2483	0.3157	0.1274
F statistic	7.8536	8.6588	2.0946
No. of observations	167	167	46

[1] Figures in parentheses are the standard errors of the coefficients
* The coefficient is significantly different from zero at the 1% level
** The coefficient is significantly different from zero at the 2.5% leve
*** The coefficient is significantly different from zero at the 5% level
§ The coefficient is significantly different from zero at the 10% leve

Table 16.   Dependent variable: <u>log</u> of HYV rice output (maunds) per acre: Bangladesh, 1975 Aman season.

(Regression coefficients)[1]

Explanatory Variables	Three villages Aggregate Data	Bogra Village
Institutional		
Log of the size of operational holding	-	.0346 (.1039)
Percentage of area leased-in.	-.0005 (.0004)	.0002 (.0010)
Hired proportion of labour-days applied per acre	-.0002 (.0004)	-.0011 (.0011)
Land Improvement Factors		
Percentage of area irrigated.	.0011§ (.0008)	.0034 (.0010)
Nitrogen fertilizer dose applied per acre (seers)	.0007 (.0004)	.0025* (.0010)
Farmyard manure applied (maunds per acre)	-.0003 (.0002)	-.0007* (.0003)
Mechanisation level in preparatory tillage (dummy)	-	-
Village Dummies (Noakhali excluded)		
Bogra	.0304 (.0325)	-
Sylhet	.0922* (.0342)	-
Intercept	.1.3633	1.3306
Mean of dependent variable	1.399	1.401
R^2	.055	.190
R^2 (adjusted)	.021	.089
F statistic	1.597	1.882
No. of observations	200	55

[1]Figures in parentheses are the standard errors of the coefficients
*The coefficient is significantly different from zero at the 1% level
**The coefficient is significantly different from zero at the 2.5% level
***The coefficient is significantly different from zero at the 5% level
§The coefficient is significantly different from zero at the 10% level

Table 17. Dependent variable: <u>log</u> of <u>local variety</u>
rice output (maunds) per acre:
Bangladesh, 1975 <u>Aman</u> season

(Regression coefficients)[1]

Data Coverage / Explanatory Variables	Three Villages Aggregate Data		Bogra Village	Sylhet Village	Noakhali Village
	Without Village Dummy	With Village Dummy			
Institutional					
Log of the size of operational holding	-0.0505*** (0.0276)	-0.0576*** (0.0289)	-0.0742** (0.0333)	-0.1078*** (0.0579)	0.2302*** (0.1067)
Percentage of area leased-in	-0.0006* (0.0003)	-0.0006** (0.0003)	-0.0004 (0.0003)	-0.0009 (0.0006)	-0.0068* (0.0014)
Hired proportion of labour-days applied	-0.0003 (0.0003)	-0.0003 (0.0003)	0.0006 (0.0004)	-0.0010§ (0.0006)	-0.0033* (0.0010)
Land Quality Determinants					
Exogenous					
Price of land (takas per acre)	0.0000 (0.0000)	0.000004** (0.00000)	0.000003 (0.00000)	0.000004 (0.00000)	0.000009*** (0.00000)
Endogenous					
Percentage of area irrigated	0.0018* (0.0006)	0.0018* (0.0066)	0.0019* (0.0006)	-	-
Nitrogen fertilizer dose applied (seers per acre)	0.0007 (0.0005)	0.0010*** (0.0005)	0.0016** (0.0006)	-	-
Farmyard manure applied (maunds per acre)	0.0007* (0.0002)	0.0007* (0.0002)	0.0006** (0.0002)	0.0008* (0.0003)	-0.0012 (0.0023)
Mechanisation level in preparatory tillage (dummy)	-0.0168 (0.0404)	-	-	-	0.0706 (0.0511)
Village Dummies (Noakhali excluded) Bogra	-	0.0510 (0.0337)	-	-	-
Syhlet	-	0.0827*** (0.0421)	-	-	-
Intercept	1.2075	1.1199	1.1347	1.2722	1.1780
Mean of dependent variable	1.2519	1.2519	1.2647	1.2382	1.2386
R²	0.1510	0.1631	0.2278	0.1841	0.5022
R² (adjusted)	0.1243	0.1334	0.1856	0.1377	0.3916
F statistic	5.6670	5.4993	5.3943	3.9702	4.5396
No. of observations	264	264	136	94	34

[1] Figures in parentheses are the standard errors of the coefficients
* The coefficient is significantly different from zero at the 1% level
** The coefficient is significantly different from zero at the 2.5% level
*** The coefficient is significantly different from zero at the 5% level
§ The coefficient is significantly different from zero at the 10% level

Table 18. Dependent variable: chemical fertiliser dose (seers) applied on an acre of HYV rice crop: Bangladesh 1974-75
(Regression coefficients)[1]

Season / Fertiliser type / Explanatory Variables	Aman 1975			Boro 1974-75		
	Nitrogen (Urea)	Potash (MP)	Phosphate (TSP)	Nitrogen (Urea)	Potash (MP)	Phosphate (TSP)
Size of operational holding (acres)	-1.582 (1.158)	2.447* (.495)	-1.266§ (.884)	-1.860** (.781)	1.953* (.405)	-0.966§ (.650)
Percentage of area leased-in	-0.129§ (.081)	—	-0.180* (.061)	-.016 (.053)	.040 (.027)	-0.039 (.044)
Membership in farmer organisation (dummy)	11.671*** (5.913)	—	18.238* (4.514)	29.063*	.595 (1.846)	23.263* (2.963)
Intercept	41.36	2.02	22.034	29.373	2.409	16.356
Mean of dependent variable	45.11	6.53	30.625	42.730	7.532	27.657
R^2	.04	.11	.12	.20	.03	.18
R^2 (adjusted)	.03	.10	.11	.19	.07	.18
F statistic	2.740	8.250	9.230	24.019	8.116	22.280
No. of observations	203	203	203	302	302	302

[1]Figures in parentheses are the standard errors of the coefficients
*The coefficient is significantly different from zero at the 1% level
**The coefficient is significantly different from zero at the 2.5% level
***The coefficient is significantly different from zero at the 5% level
§The coefficient is significantly different from zero at the 10% level

Table 19. Dependent variable: number of times weeding operation
applied to <u>rice crops</u>: Bangladesh 1974-75
(Regression coefficients)[1]

Explanatory Variables	Boro 1974-75		Aman 1975
	HYV	Traditional	Traditional
Size of operational holding (acres)	-.029§ (.018)	-.035*** (.018)	--
Percentage of area leased-in	-.001 (.001)	-.003*** (.001)	-.0022§ (.0013)
Membership in farmer organisation (dummy)	.335* (.082)	--	.222* (.086)
Intercept	2.186	1.116	.551
Mean of dependent variable	1.883	.970	.629
R^2	.07	.04	.05
R^2 (adjusted)	.06	.03	.04
F statistic	5.193	3.567	5.172
No. of observations	299	167	237

[1]Figures in parentheses are the standard errors of the coefficients

*The coefficient is significantly different from zero at the 1% level

**The coefficient is significantly different from zero at the 2.5% level

***The coefficient is significantly different from zero at the 5% level

§The coefficient is significantly different from zero at the 10% level

Table 20. Dependent variable: Cropping Intensity among Adopters and Non-Adopters of HYV Rice, Bogra, Noakhali and Sylhet Villages of Bangladesh, 1976 (Regression coefficients)[1]

Season, HYV Adaption Status / Explanatory Variables	BORO SEASON				AMAN SEASON			
	HYV ADOPTERS		HYV NON-ADOPTERS		HYV ADOPTERS		HYV NON-ADOPTERS	
	Without Village Dummy	With Village Dummy	Without Village Dummy	With Village Dummy	Without Village Dummy	With Village Dummy	Without Village Dummy	With Village Dummy
Institutional								
Size of operational holding (acres)	-6.0659* (1.1672)	-3.5831* (1.0298)	-2.6143 (7.3424)	-	-3.6560* (1.3449)	-2.7597*** (1.3955)	-8.6787* (2.2738)	-4.3605** (1.8603)
Percentage of area leased-in	-0.0441 (0.0905)	0.0555 (0.0733)	-0.3237 (0.3214)	-0.2948 (0.3418)	-0.0190 (0.1100)	0.0569 (0.0975)	-0.1027 (0.1529)	0.0121 (0.1199)
Membership in farmers organisation (dummy)	25.3007* (6.2690)	-	-7.0943 (22.7742)	-16.3958 (26.7968)	10.5589 (8.5264)	-	23.4695* (10.3199)	4.7628 (8.4072)
Technical Percentage of area irrigated during:								
AMAN Season	0.2867§ (0.1747)	0.0238 (0.1395)	3.8327* (1.6214)	3.8226*** (1.8020)	0.3248 (0.1984)	0.1041 (0.1794)	0.3390 (0.3490)	0.1295 (0.2729)
BORO Season	0.1599§ (0.0927)	-0.0196 (0.0746)	-	0.0737 (0.3608)	0.0266 (0.0943)	-0.0779 (0.0832)	0.3375** (0.1517)	0.1727 (0.1200)
Village Dummies (Noakhali excluded)								
BOGRA	-	35.8723* (6.6681)	-	52.2464 (57.6190)	-	33.5599* (9.4592)	-	78.1235* (10.4514)
SYLHET	-	-43.0040* (5.3684)	-	-	-	-33.5370* (7.2939)	-	-
Intercept	158.659	196.721	168.741	164.719	179.830	198.312	146.372	136.591
Mean of dependent variable	176.679	176.679	158.402	158.402	184.381	184.381	158.926	158.926
R^2	0.184	0.499	0.353	0.392	0.072	0.303	0.218	0.534
R^2 (adjusted)	0.165	0.485	0.181	0.111	0.042	0.276	0.172	0.499
F statistic	10.079	37.002	2.048	1.395	2.404	11.111	4.685	15.767
No. of observations	230	230	20	20	160	160	90	90

[1]Figures in parentheses are the standard errors of the coefficients

*The coefficient is significantly different from zero at the 1% level

**The coefficient is significantly different from zero at the 2.5% level

***The coefficient is significantly different from zero at the 5% level

§The coefficient is significantly different from zero at the 10% level

Chapter 4

Labour Utilisation Under Technical Change

Introduction

In this chapter an attempt will be made to identify and
assess the determinants of the intensity of labour use in rice
cultivation. Before embarking on an econometric analysis of
the cross-sectional survey data, a brief description of the
rough theoretical framework is presented and the nature of
hypotheses being tested are discussed.[1] Several positive yield-
increasing or land improvement factors like irrigation and fertilization could be identified as significantly influencing the
farm labour day per acre of cultivated area. Land improvement
factors could be both labour-using or labour-saving depending on
their specific nature. For example, lift irrigation is normally
more labour-using than gravity flow irrigation. Farmyard manure
is more labour-using than chemical fertilizers. Use of fertilizers
apart from increasing yields, also increases the growth of weeds,
thereby increasing the need for per acre labour input. Improved
farm practices could be labour-intensive. Transplanting of paddy
in rows facilitates the weeding operation and thereby increases
the labour input for weeding. The level of mechanization in preparatory tillage would also be an important determinant of the
intensity of per acre labour-use.

Having reviewed the possible direction in which the intensity
of labour use per acre may alter as a result of changes in technical
inputs, the likely impact of some important institutional variables
on labour-use are discussed. Land tenancy, the first institutional
factor, may depress productivity and labour use on account of the

---

[1]A similar framework was developed and applied by Bardhan to
the data from India. For the mathematical derivation and elaboration
of the model, see Bardhan [10].

tenurial insecurity as well as the prevalent crop-sharing system.[1]
The second institutional factor considered is farm size. The
theoretical framework suggests that as farm size increases labour
intensity per acre decreases under both diminishing and constant
returns to scale.[2] Furthermore, a large body of empirical studies
for India have established that small farms put in more labour input per acre.[3] For smaller farms, dependent largely on family
labour, the imputed cost of labour is less than that for the
larger farms hiring labour at the market wage rate. Another reason
why labour intensity is higher on family-based smaller farms is
better quality of management that can be applied; while larger
farms have to incur supervision costs for hired labour. Moreover,
the need for small farmer's survival drives him to more intensive
effort. A poor peasant family, depending on a small piece of land,
would not only ignore any marginal productivity calculations insofar
as family labour is concerned, he would employ hired labour whenever necessary to supplement the family labour and in so doing
would pay no heed to marginal productivities. He would try to
improve the quality of the land by small-scale irrigation, higher
frequency of weeding and other such means as can be procured with
the help of labour.

---

[1]The traditional viewpoint is that share-cropping causes inefficient allocation of resources because, unless landowner and
sharecropper share all input costs in the same proportion as output,
the sharecropper will apply below optimal levels of variable inputs
since he will equate the full marginal cost of such inputs with
his share of their marginal product, not with the total marginal
product. The previous chapter discusses further development of this
theory. As indicated earlier, given that sharecropping is by and
large the only tenancy system that prevails in Bangladesh agriculture,
some negative effects of tenancy on labour input should be observed.

[2]The empirical studies on returns to scale in developing country
agriculture generally have found approximately constant returns to
scale [13]. As we have already noted from the previous chapter that
the returns to scale in Bangladesh agriculture is not significantly
different from constant returns and this feature is equally valid for
the traditional and HYV rice crops. An important theoretical point is
that the shifting relative price of land and labour induces the larger
farms to use relatively less labour and more land regardless of returns to scale in the technical production function.

[3]For a review and survey of this evidence see Chattopadhyay and
Rudra [20].

However, based on data available for various regions in India, reservations have been expressed about the universal validity of the inverse relationship between intensity of labour use and farm size [20]. Another note of caution has been delivered on the effect of aggregation, or pooling of data pertaining to, different villages. In such a situation there is a risk of falsifying the dependence of the intensity of labour-use on farm size in both directions [20]. The statistical error on account of the pooling of data from different villages could result in either a spurious inverse relationship or the intra-village inverse relationship being obliterated.[1]

It can also be expected that the total labour input per acre would depend on the mode of labour use. That is to say, farms employing exclusively family labour use labour more intensively than farms based exclusively on hired labour. Finally, it is being postulated that the use of hired labour is dependent on farm size, land tenancy, the number of adult farm workers in the household, the proportion of rice area under HYV and the level of mechanisation used for preparatory tillage.

Hypotheses

To sum up, the following hypotheses are being tested. Labour intensity per acre is expected to be _positively_ associated with the use of lift irrigation (compared to gravity flow irrigation), with fertilization (unless farmyard manure is largely substituted by chemical fertilizers), with the practice of row planting of paddy and with the frequency of weeding; it is expected to be _negatively_ associated with the use of mechanisation for preparatory tillage and with institutional variables like farm size, land tenancy and proportion of hired labour to total labour use. Use of hired labour is expected to be _positively_ associated with the proportion of area under HYV and with the use of mechanisation for preparatory tillage;

---

[1]Although this statistical point there was stressed with respect to the dependence of productivity on farm size, it would be equally applicable to the analysis of the relationship between the intensity of labour use and farm size. Reasons for the emergence of this error is provided with an illustration in the previous chapter.

it is expected to be <u>positively</u> associated with farm size, with
land tenancy and with the number of adult farm workers in the
household.

It is of major policy interest to examine whether the introduction of the green revolution technology has altered (weakened
or strengthened) the inverse relationship between the per acre
intensity of labour use and farm size and that with land tenancy.
No empirical evidence appears to be available to shed light on
this relationship following technological change. The few observations that have been made pertain to the relationships between productivity per acre and farm size after the introduction of the new
technology and these statements are highly speculative and conjectural rather than being based on any rigorous and systematic treatment of data [13].

In the following section, we proceed with the empirical evaluation of these hypotheses with the help of multiple regression
analysis of farm-level cross-sectional variations in labour intensity
per acre (for total, operational and hired labour separately) for
both the Aman and Boro seasons using both pooled (aggregate) and
intra-village (disaggregate) data. We begin with the analysis of
the level and determinants of total labour use per acre.

Use of total labour days

The average number of farm labour days (in terms of standardised
8-hour man days) used per acre of cropped area under HYV rice is
115.5 days during the Boro (winter) season as compared to 81 days
for the local rice variety. During the Aman (summer) season these
are respectively 76 and 62 labour days. These average figures point
to the wide differences in the intensity of labour use with variations in crop variety and season. Tables 21-24 present the results
of a linear regression analysis of the determinants of total mandays
applied per acre of cropped area for both seasons and crop varieties.
Total mandays applied per acre during the Boro season are positively

associated with irrigation; and they are negatively associated with
the level of mechanisation. With the introduction of the green
revolution technology we observe the emergence and strengthening
of the inverse relationship between the per acre intensity of
labour-use and farm size; and this is established in terms of
both pooled and intra-village data and appears to be quite universal (Table 21). In marked contrast, this relationship for
farms growing local rice varieties appears to be positive in terms
of aggregate data but no significant relationship is observed
when intra-village data are analysed (Table 22). Land tenancy has
a depressing effect on the intensity of labour-use following the
introduction of the green revolution technology, and this too is
established by both pooled and intra-village data although it is
not universal (Table 21). In contrast, no significant relationship
is observed between labour intensity and land tenancy on local rice
variety farms (Table 22). No significant relationship between the
intensity of labour use and the hired proportion of labour-days is
observed.

There are wide regional differences in the intensity of labour
use during the Boro season. For example, the labour days applied
by HYV farms in Bogra is 68 days more than those belonging to
Noakhali (Table 21). One reason is the concentration of highly
labour-intensive hand tubewells in Borgra as compared to the
Noakhali village which relies almost exclusively on power pump
(gravity flow) irrigation which is relatively less labour-using.

During the Aman (summer season) intensity of labour use is
positively associated with land improvement factors like fertilization (primarily through the use of farmyard manure), with irrigation
and with the frequency of weeding; it is negatively associated with
the level of mechanization for preparatory tillage (Tables 23 and 24).
Once again it is observed that the inverse relationship between the
intensity of labour use and farm size has emerged following the
introduction of the green revolution technology during the Aman

season (Table 23). It is established in terms of both pooled
and intra-village data although the relationship does not appear
to be completely universal. This is in contrast to the local rice
variety farms where this relationship is found to be positive with
aggregate data and is not significant when intra-village data is
considered (Table 24). Land tenancy does not appear to have any
significant depressing effect on the intensity of labour-use.
Labour intensity does not appear to be related to the mode of
labour use during Aman, although with one exception for the Bogra
village where a significant positive association is observed. This
may be on account of the fact that this village is marked by a high
rate of landlessness which leads to a large body of hired labour.

Regional differences in labour intensity do not appear to be
important during the Aman season. Only the Sylhet village applies
significantly higher labour input per acre of local variety rice
compared to that for the Noakhali village.

Labour Input by Farm Operations

Having looked at the determinants of total labour use per acre
for rice cultivation, an attempt is made below to analyse the determinants of operation-wise variations in labour-intensity. Much
greater insight is likely to be obtained of the impact of individual
technical factors and institutional variables through this approach.
The first farm operation is preparatory tillage, which is discussed
below.

Preparatory Tillage Labour Days

Tillage labour days per acre of cropped area are on an average
22 days for both HYV and local paddy during the Boro season as
compared to 18 and 17 days respectively for the HYV and local
varieties during the Aman season (Tables 25 to 28). This implies
that not much difference in labour input for preparatory tillage
exists between crop varieties and between seasons.

Tillage labour days are negatively associated with the level of mechanisation for preparatory tillage and it is statistically significant for HYV farms both during the Aman and Boro seasons (Tables 25 and 27).[1] The use of tractor/power tiller on an acre of HYV plot leads to the displacement of 4 to 6 labour days. The inverse relationship between tillage labour days and farm size remains unaltered with the introduction of the new technology during the Boro season and this is established both in terms of pooled and intra-village data, although it does not appear to be universal (Tables 25 and 26). The emergence of this relationship is observed after the introduction of the green revolution technology during the Aman season and this is based on both pooled and intra-village data, although it is not universal (Table 27). Interestingly enough no such relationship is observed for local variety farms during Aman. Land tenancy has a depressing effect on tillage labour days used only after the introduction of the green revolution during Boro (Table 25).

Since the use of tractors/power tillers is concentrated in the Noakhali village, it is little surprising that tillage labour input in the Sylhet village is significantly higher than that of Noakhali. No significant difference was observed for the Bogra village.

Labour Use for Fertilization

The average fertilization labour days applied per acre ranges from 1.6 to nearly 4 labour days (Tables 29 to 32).

Fertilizer dose applied is a significant determinant of labour days used for all crops. Farmyard manure is labour-using; and nitrogen (Urea) fertilizer is the only chemical fertilizer which also seems to be labour-using. Substitution of other chemical fertilizers for farmyard manure has had a labour-saving impact, although not consistently.

---

[1]A total of 58 bullock-tractor comparisons (19 tested statistically) in South Asia reveals that the use of a tractor is associated neither with an increase nor a decrease in labour use per unit of land, although evidence may slightly favour a decreasing effect [15].

The inverse relationship between fertilisation labour days
and farm size has emerged following the introduction of the green
revolution technology during the Boro season and this is established
in terms of both pooled and intra-village data; although this relationship is not completely universal (Tables 29 and 30). However,
this relationship is observed to be significant for the local
variety Boro crop only in terms of pooled data (Table 30). This
relationship was not found to be significant for either crop
during the Aman season (Tables 31 and 32).

No significant relationship between tenancy and labour use
in this operation is observed for any crop.

Regional differences in labour input for fertilization have
been found to be significant.

Weeding labour days

Per acre average labour days for weeding were 24.3 and 13.2
days respectively for HYV and local variety rice during the Boro
season (Tables 33 and 34). These were respectively 13.3 and 5.4
man-days during the Aman season (Tables 35 and 36). This reflects
the wide seasonal and varietal differences in the use of labour for
weeding.

As expected per acre labour days for weeding are by and large
positively associated with the dose of fertilizer/farmyard manure
applied although it is not consistent for all crops and seasons.
Labour intensity in this operation is positively associated with
the frequency of weeding for all crops. Weeding labour input appears
to have a significant positive association with row planting prac-
tised during the Aman season (in terms of pooled data only).

The inverse relationship between weeding labour input and farm
size is observed during the Boro season in terms of pooled data for
local rice variety (Table 34). During Aman the relationship is

similarly observed for local variety crop (in terms of both pooled
and intra-village data). For none of these crops does this relationship appear to be universal.

The depressing effect of land tenancy on labour intensity
for weeding is observed only in the Bogra village in respect of
HYV Boro and the local variety Aman rice (Tables 33 and 36).

The Bogra village appears to be applying significantly less
labour days for weeding as compared to that of Noakhali; the Sylhet
village appears to be using significantly more compared to Noakhali.

Irrigation Labour Days

Average labour days applied for irrigating an acre of HYV rice
field is 20.3 days as compared to only 3.8 days for the local rice
variety (Tables 37 and 38). Method of irrigation used is a significant determinant of the intensity of labour use in this operation.
As expected irrigation labour input per acre is positively associated with hand tubewell, powered tubewell and indigenous methods
whereas the power pump (gravity-flow) irrigation was not significant.

It is most noteworthy that the emergence of the inverse relationship between labour intensity for irrigation and farm size is observed after the introduction of the green revolution technology
(Table 37). It is quite natural, the small farmer should try to
improve the quality of his land by small-scale irrigation (such as
hand tubewell) with the help of labour. It is equally noteworthy
that land tenancy depresses the intensity of labour use for irrigation after the introduction of the green revolution technology.

The Bogra village where the bulk of the hand tubewells are
concentrated applies significantly higher labour days for irrigating
HYV plots as compared to that of Noakhali. For irrigating traditional rice varieties Sylhet applies significantly higher labour
input compared to that of Noakhali, probably because Sylhet uses
more of the indigenous irrigation methods near the "haor" (marshy
land) areas.

Use of Hired Labour

The inter-farm variations in total and operation wise labour-use did not capture the inter-farm variations in the use of hired labour. Therefore, in the following section an attempt is made to identify the determinants of hired labour use.

Hired mandays

On an average 27.3 hired labour days are applied per acre during the Boro season as compared to 18.2 days during the Aman season (Tables 39 and 41). As expected hired labour days applied per acre are positively associated with the proportion of area under HYV (with the only exception of the Noakhali village which shows a negative association for the Aman season but is not significant). There does not appear to be any association between the use of hired labour days with the level of mechanisation for preparatory tillage.

As expected hired labour days applied per acre are positively associated with farm size both for pooled and intra-village data and it is almost completely universal.

Hired labour days applied per acre are negatively associated with land tenancy and with the number of adult agricultural workers in the household. This relationship has been established both in terms of pooled and intra-village data and its validity is virtually universal.

Only during the Aman season do the Bogra and Sylhet villages apply significantly higher hired labour days per acre compared to Noakhali village.

Hired proportion of total labour use

Having examined the determinants of the absolute number of labour days applied per acre it is of interest to look at the inter-farm variations in the hired proportion of labur days applied per acre. During the Boro and Aman seasons on an average 51.7 per cent and 51.3 per cent of the respective total labour applied per acre are hired (Tables 40 and 42).

During the Boro season hired proportion of total labour days applied per acre is <u>positively</u> associated with the percentage of rice under HYV (excepting Bogra village which is not significant). During Aman this relationship is found to be significant only for the pooled data. It is <u>positively</u> associated with the level of mechanisation used for preparatory tillage only for pooled data during both seasons.

Hired proportion of total farm days applied per acre is <u>positively</u> associated with farm size and this is firmly established in terms of both pooled and intra-village data; and this result is universal.

It is <u>negatively</u> associated with land tenancy and with the number of adult agricultural workers in the household and this relationship is established both in terms of pooled and intra-village data; it is universal (with the only exception of the Bogra village during the Boro season).

Hired proportion of total labour applied per acre in the Sylhet district appears to be significantly less than that of the Noakhali village.

Summary and Conclusions

The above analysis of the inter-farm variations in labour use shows that the per acre labour intensity is <u>positively</u> associated with land improvement factors like irrigation, fertilization and weeding; it is <u>negatively</u> associated with the use of mechanisation for preparatory tillage. The analysis of the operation-wise variations in labour intensity is more revealing. Per acre labour

intensity is positively associated with the dose of farmyard
manure and nitrogen (urea) fertilizer applied per acre. Substitution of the other types of chemical fertilizers (phosphate
and potash) for farmyard manure is labour-saving. Labour input
for weeding is positively associated with the fertilizer/farmyard
manure dose applied, with the frequency of weeding and with the
practice of row planting. Irrigation method used is a significant
determinant of the intensity of labour-use in this operation. As
expected irrigation labour input per acre is positively associated
with the use of hand tubewells, powered tubewells and indigenous
methods whereas the power pump (gravity flow) irrigation was not
significant.

Use of mechanisation for preparatory tillage leads to the
loss of 4 to 6 labour days of employment per acre and the average
labour intensity appears to be invariant to crop variety and
agricultural season.

With the introduction of the green revolution technology the
inverse relationship between per acre labour intensity and farm
size is observed and this relationship is established in terms of
pooled and intra-village (disaggregate) data. Although this does
not appear to be a fully universal law, this feature occurs repeatedly in different combinations of crop varieties, seasons and
specific farm operations in different parts of the country. For
example, the inverse relationship between labour intensity for
preparatory tillage and farm size remains unaltered with the introduction of the green revolution technology during the Boro season
in respect of pooled data and Noakhali village. However, the
emergence of this inverse relationship in this operation is observed
during the Aman season (also in respect of pooled data and Noakhali
village). Similar inverse relationship is observed for fertilization, weeding and irrigation operations again in different combinations of crop varieties, seasons and regions.

With the introduction of the green revolution technology
the emergence of a significant depressing effect of land tenancy
on the intensity of labour use per acre is observed. This is
established in terms of both pooled and intra-farm data. As in
the case of the inverse relationship between labour intensity and
farm size, this is not a fully universal law, although this
feature occurs in different combinations of crop variety, season
and specific farm operation in different parts of the country.
For example, with the introduction of the green revolution technology a significant depressing effect of land tenancy on labour
intensity is observed for only one crop (Boro HYV in Noakhali
village).

Dependence of labour intensity on the mode of labour use
does not, by and large, appear to be significant with the exception of the Bogra village during Aman and the pooled data in
respect of HYV Aman.

Over one half of the entire labour days applied on an acre
of rice crop cultivated is hired labour. Hired labour (number
of labour days as well as hired proportion of total labour) applied
per acre is positively associated with the percentage of the cropped
area under HYV. Per acre use of hired labour days does not depend
upon the level of mechanisation in preparatory tillage, although
the hired proportion of total labour use is positively related to
it only in respect of pooled data.

Hired labour (number of labour days as well as hired proportion of total labour) applied per acre is positively associated
with farm size and this is convincingly established in terms of
pooled and intra-village (disaggregate) data; and this is a totally
universal phenomenon.

Hired labour (number of labour days as well as hired proportion
of total labour) applied per acre is negatively associated with land
tenancy and this is established in terms of pooled and intra-village

data and this feature is virtually universal (with the only exception of one crop grown in Bogra village).

Hired labour applied per acre is <u>negatively</u> associated with the number of adult agricultural workers in the household and this is also established in terms of pooled and intra-village data; it is also universal (with the only exception of the Boro rice crop in Bogra).

Significant regional differences in the per acre intensity of labour use is observed. In general, per acre labour intensity in the Bogra and Sylhet villages are higher than that of Noakhali. However, slight variations to this feature are to be found in different combinations of crops, seasons and specific farm operations. This phenomenon is also reflected in a significantly higher intensity of hired labour days use in the Bogra and Sylhet villages. However, hired proportion of total labour days applied per acre in Sylhet is significantly less than that of Noakhali.

Table 21. Dependent variable:
Mandays applied per acre
of cropped area under
HYV rice: Bangladesh
1974-75 Boro Season.

(Regression coefficients)[1]

Data Coverage / Explanatory Variables	Three Villages Aggregate Data		Bogra Village	Sylhet Village	Noakhali Village
	Without Village Dummy	With Village Dummy			
Institutional					
Size of operational holding (acres)	-1.6405 (1.3542)	-3.7518* (1.3968)	-13.4402* (5.3393)	0.6181 (1.5298)	-6.8478* (2.4352)
Percentage of area leased-in	-0.1405 (0.0926)	-0.2076* (0.0911)	-0.7001 (0.5171)	-0.0941 (0.1072)	-0.2343* (0.0823)
Hired proportion of labour days applied per acre	-0.1562 (0.1060)	-0.1109 (0.1028)	0.4299 (0.4796)	-0.3232* (0.1297)	---
Technical					
Mechanisation level in preparatory tillage (dummy)	-13.9784§ (8.6943)	---	---	-18.7556 (23.4779)	1.7753 (4.7864)
Row-planting practised (dummy)	21.1247§ (11.9157)	28.8257* (11.8848)	49.2351 (33.2837)	-24.7882 (18.1874)	27.5474 (22.5915)
Source of irrigation water (percentage of irrigated area under:)					
Power pump	-0.0299 (0.0703)	0.1508*** (0.0788)	---	0.1467§ (0.0865)	0.1038§ (0.0676)
Powered tubewell	0.2249* (0.0820)	0.0735 (0.1127)	0.0610 (0.2097)	---	---
Hand tubewell	1.7486* (0.1312)	1.5048* (0.1857)	1.6622* (0.3487)	---	---
Indigeneous	0.1421* (0.0426)	0.1264* (0.0420)	0.4191 (0.3148)	0.1015* (0.0395)	0.4929 (1.1301)
Village Dummies (Noakhali excluded)					
Bogra	---	68.1722* (18.8940)	---	---	---
Sylhet	---	37.3551* (8.4064)	---	---	---
Intercept	96.5744	55.5555	105.0521	142.4518	58.3980
Mean of dependent variable	115.5887	115.5887	195.1268	114.2039	86.2108
R^2	0.4766	0.5136	0.5980	0.1010	0.1500
R^2 (adjusted)	0.4591	0.4955	0.5043	0.0554	0.0927
F statistic	27.3136	28.4016	6.3764	2.2150	2.6182
No. of observations	280	280	186	146	96

[1]Figures in parentheses are the standard errors of the coefficients
*The coefficient is significantly different from zero at the 1% level
**The coefficient is significantly different from zero at the 2.5% level
***The coefficient is significantly different from zero at the 5% level
§The coefficient is significantly different from zero at the 10% level.

Table 22. Dependent variable: Mandays applied per acre of cropped area under local rice variety: Bangladesh, 1974-75 Boro Season

(Regression coefficients)[1]

Data coverage Explanatory Variables	Three Villages Aggregate Data Without Village Dummy	With Village Dummy	Sylhet Village	Noakhali Village
Institutional				
Size of operational holding (acres)	3.6530* (1.3965)	1.0533 (1.3759)	2.2206 (1.5709)	-8.6284* (3.0477)
Percentage of area leased-in	0.0852 (0.1039)	0.0169 (0.0976)	-0.0343 (0.1168)	-0.1975§ (0.1415)
Hired proportion of labour-days applied per acre	-0.1847 (0.1183)	-0.1759§ (0.1093)	-0.3805* (0.1518)	0.1006 (0.0992)
Technical				
Mechanisation level in preparatory tillage (dummy)	-21.3759** (9.6424)	-4.1878 (9.6636)	-40.7487 (27.2088)	1.3523 (5.5347)
Row planting practised (dummy)	-4.0601 (11.5419)	--	--	17.1846 (17.5801)
Source of irrigation water (proportion of irrigated area under:)				
Power pump	-0.0545 (0.0630)	0.2097* (0.0673)	0.3301* (0.1006)	0.1509* (0.0527)
Powered tube well	0.1538 (0.1249)	0.9393** (0.4393)	--	--
Hand tube well	--	--	--	--
Indigeneous	0.1758* (0.0352)	0.1477* (0.0331)	0.1652* (0.0371)	--
Village Dummies (Noakhali excluded)				
Bogra	--	-100.98 (78.50)	--	--
Sylhet	--	45.8096* (9.6689)	--	--
Intercept	71.6566	35.0617	83.4224	24.8236
Mean of dependent variable	80.8957	80.8957	93.4278	49.0308
R^2	0.2583	0.3688	0.2222	0.3153
R^2 (adjusted)	0.2174	0.3293	0.1760	0.2011
F statistic	6.3126	9.3467	4.8088	2.7624
No. of observations	154	154	108	43

[1] Figures in parentheses are the standard errors of the coefficients
*The coefficient is significantly different from zero at the 1% level
**The coefficient is significantly different from zero at the 2.5% level
***The coefficient is significantly different from zero at the 5% level
§The coefficient is significantly different from zero at the 10% level

Table23. Dependent variable: man-days applied per
acre of cropped area under <u>HYV rice</u>:
Bangladesh, 1975 Aman.

(Regression coefficients)[1]

Explanatory Variables	Data Coverage — Three Villages Aggregate Data		Bogra Village	Sylhet Village	Noakhali Village
	Without Village Dummy	With Village Dummy			
Institutional					
Size of operational holding (acres)	-0.4967 (0.9697)	-2.1447*** (1.1044)	-4.5981*** (2.2706)	-0.5255 (1.5671)	-5.1738** (2.2206)
Percentage of area leased-in	0.0066 (0.0706)	-0.0317 (0.0705)	-0.1449 (0.1733)	-0.1555§ (0.1099)	-0.0909 (0.0882)
Hired proportion of labour days applied per acre	0.0976 (0.0681)	0.1532*** (0.0693)	-0.4725* (0.1780)	-0.0274 (0.1181)	0.0689 (0.0833)
Technical					
Mechanisation level in preparatory tillage (dummy)	-17.8621* (5.0750)	-12.1447** (5.3406)	-	-7.5206 (16.2238)	-8.8692*** (4.6428)
Percentage of cultivated area irrigated	0.1689 (0.1209)	0.0742 (0.1244)	0.0643 (0.2748)	-0.0701 (0.2301)	0.0832 (0.1561)
Nitrogen fertilizer (urea) applied (seers per acre)	0.0442 (0.0590)	0.0607 (0.0608)	0.3166§ (0.1660)	0.1443 (0.1382)	-0.0224 (0.0610)
Farmyard manure applied (maunds per acre)	0.0560** (0.0256)	0.0314 (0.0264)	-0.0641 (0.0477)	0.1190* (0.0432)	0.0981 (0.1196)
Village Dummies (Noakhali excluded)					
Bogra	-	16.5883* (6.4937)	-	-	-
Sylhet	-	16.7577* (6.0361)	-	-	-
Intercept	68.920	61.375	59.097	76.639	72.369
Mean of dependent variable	75.661	75.661	86.354	81.123	67.697
R^2	0.125	0.165	0.218	0.292	0.142
R^2 (adjusted)	0.093	0.126	0.123	0.146	0.079
F statistic	3.956	4.211	2.281	2.001	2.264
No. of observations	202	202	56	42	104

[1] Figures in parentheses are the standard errors of the coefficients
* The coefficient is significantly different from zero at the 1% level
** The coefficient is significantly different from zero at the 2.5% level
*** The coefficient is significantly different from zero at the 5% level
§ The coefficient is significantly different from zero at the 10% level

Table 24. Dependent variable: Mandays applied per acre
of cropped area under local rice variety:
Bangladesh, 1975 Aman season.

(Regression coefficients)[1]

Data coverage / Explanatory Variables	Three Villages Aggregate Data		Bogra Village	Sylhet Village	Noakhali Village
	Without Village Dummy	With Village Dummy			
Institutional					
Size of operational holding (acres)	1.2027§ (.6395)	.1487 (.6728)	-.1147 (.7408)	.2878 (1.2295)	-3.1120 (3.6064)
Percentage of area leased-in	-.0094 (.0450)	.0222 (.0437)	-.3480 (.0421)	-.0124 (.0928)	-.1485 (.2099)
Hired proportion of labour-days applied per acre.	-.0179 (.0482)	.0101 (.0472)	.1011*** (.0503)	-.0824 (.0985)	-.1095 (.1400)
Technical					
Mechanisation level in preparatory tillage (dummy).	-5.8014 (5.4550)	-2.3468 (5.3525)	-	-13.8388 (26.2422)	1.4061 (6.9453)
Percentage of cultivated area irrigated.	.1976** (.0895)	.2272* (.0870)	.3309* (.0752)	-.0663 (.2132)	-
No. of times weeding practised	7.6100* (2.1706)	6.5091* (2.1199)	.5002 (2.270)	16.4727* (4.4315)	-1.8437 (8.3296)
Farmyard manure applied (maunds per acre)	.0953* (.0249)	.0832* (.0243)	.0703* (.0262)	.0662** (.0457)	.6998*** (.3207)
Village Dummies (Noakhali excluded)					
Bogra	-	-	-	-	-
Sylhet	-	12.6665* (3.1616)	-	-	-
Intercept	51.6388	49.411	48.4549	60.3131	63.9052
Mean of dependent variable	62.1283	62.1283	56.9193	71.9252	57.1662
R^2	.155	.211	.247	.212	.1966
R^2 (adjusted)	.129	.183	.206	.136	.0244
F statistic	5.948	7.556	6.107	2.800	1.1419
No. of observations	235	235	119	81	35

[1]Figures in parentheses are the standard errors of the coefficients.
*The coefficient is significantly different from zero at the 1% level
**The coefficient is significantly different from zero at the 2.5% level
***The coefficient is significantly different from zero at the 5% level
§ The coefficient is significantly different from zero at the 10% level

Table 25. Dependent variable: Preparatory tillage labour days per acre of cropped area under <u>HYV rice</u>: Bangladesh 1974-75 Boro season.

(Regression coefficients)[1]

Explanatory Variables	Three Villages Aggregate Data	Noakhali Village
Institutional		
Size of operational holding (acres)	-.6695 *** (.3250)	-3.0921 * (.8839)
Percentage of area leased-in.	-.0316 § (.0207)	-.0496 § (.0303)
Technical		
Mechanisation level in preparatory tillage (dummy)	-4.3261 *** (2.2537)	-3.8383 ** (1.7394)
Village Dummies (Noakhali excluded)		
Bogra	2.6441 § (2.4177)	-
Sylhet	13.3656 (1.8079)	-
Intercept	17.1721	19.9734
Mean of dependent variable	22.0446	14.3040
R^2	.257	.185
R^2 (adjusted)	.244	.159
F statistic	20.477	7.118
No. of observations	302	98

[1] Figures in parentheses are the standard errors of the coefficients.
*The coefficient is significantly different from zero at the 1% level.
**The coefficient is significantly different from zero at the 2.5% level.
***The coefficient is significantly different from zero at the 5% level.
§The coefficient is significantly different from zero at the 10% level.

Table 26.   Dependent variable: Preparatory tillage labour
days per acre of cropped area under local rice
variety: Bangladesh, 1974-75 Boro season.

(Regression coefficients)[1]

Data coverage Explanatory Variables	Three Villages Aggregate Data	Noakhali Village
Institutional		
Size of operational holding (acres)	-1.1320* (.4342)	-2.7050** (1.1263)
Percentage of area leased-in	-.0087 (.0298)	-.0648 (.0557)
Technical		
Mechanisation level in preparatory tillage (dummy)	-3.9835 (3.1676)	-2.5376 (2.2602)
Village Dummies (Noakhali excluded)		
Bogra	1.3476 (6.2684)	-
Sylhet	13.2882 (2.5512)	-
Intercept	16.6940	18.7474
Mean of dependent variable	22.3924	13.4815
R^2	.235	.184
R^2 (adjusted)	.212	.131
F statistic	10.4370	3.4522
No. of observations	176	50

[1]Figures in parentheses are the standard errors of the coefficients.
*The coefficient is significantly different from zero at the 1% level.
**The coefficient is significantly different from zero at the 2.5% level.
***The coefficient is significantly different from zero at the 5% level.
§The coefficient is significantly different from zero at the 10% level.

Table 27. Dependent variable: Preparatory tillage
labour days per acre of cropped area under
HYV rice: Bangladesh, 1975 Aman.

(Regression coefficients)[1]

Explanatory Variables	Three Villages Aggregate Data	Noakhali Village
Institutional		
Size of operational holding (acres)	-.9418* (.3913)	-2.6768* (.9681)
Percentage of area leased-in	-.0336§ (.0251)	-.0348 (.0375)
Technical		
Mechanisation level in preparatory tillage (dummy)	-5.9505* (1.9838)	-5.7932* (1.9142)
Village Dummies (Noakhali excluded)		
Bogra	3.4773§ (2.0076)	-
Sylhet	6.4203 (2.2301)	-
Intercept	19.6378	21.4991
Mean of dependent variable	18.2835	15.9288
R^2	.129	.176
R^2 (adjusted)	.107	.151
F statistic	5.8482	7.1082
No. of observations	203	104

[1]Figures in parentheses are the standard errors of the coefficients.
*The coefficient is significantly different from zero at the 1% level.
**The coefficient is significantly different from zero at the 2.5% level.
***The coefficient is significantly different from zero at the 5% level.
§The coefficient is significantly different from zero at the 10% level.

Table 28. Dependent variable: Preparatory tillage
labour days per acre of cropped area under
local rice variety: Bangladesh, 1975 Aman.

(Regression coefficients)[1]

Explanatory Variables	Three Villages Aggregate Data
Institutional	
Size of operational holding (acres)	-
Percentage of area leased-in	-.0128 (.0165)
Technical	
Mechanisation level in preparatory tillage (dummy)	-3.7710 (2.8136)
Village Dummies (Noakhali excluded)	
Bogra	-.6055 (2.1780)
Sylhet	6.7039* (2.2047)
Intercept	15.4017
Mean of dependent variable	16.8890
R^2	.154
R^2 (adjusted)	.141
F statistic	12.047
No. of observations	270

[1] Figures in parentheses are the standard errors of the coefficients.
* The coefficient is significantly different from zero at the 1% level.
** The coefficient is significantly different from zero at the 2.5% level.
*** The coefficient is significantly different from zero at the 5% level.
§ The coefficient is significantly different from zero at the 10% level.

Table 29. Dependent variable: mandays per acre required
for fertiliser application under <u>HYV rice</u>:
Bangladesh, 1974-75 Boro season

(Regression coefficients)[1]

Data coverage / Explanatory Variables	Three Villages Aggregate Data		Sylhet Village	Noakhali Village
	Without Village Dummy	With Village Dummy		
Institutional				
Size of operational holding (acres)	-.3233* (.0978)	-.2413** (.1003)	-.2175* (.0585)	-.6303 § (.3777)
Percentage of area leased-in	-.0055 (.0062)	-.0030 (.0062)	-.0054 (.0038)	-.0045 (.0130)
Technical				
Fertiliser dose applied (seers per acre)				
Nitrogen (urea)	.0351* (.0083)	.0295* (.0089)	.0290* (.0083)	.0294 *** (.0148)
Phosphate (TSP)	.0141 (.0097)	-.0065 (.0111)	.0325* (.0100)	-.0346 § (.0178)
Potash (MP)	-.0288*** (.0144)	.0090 (.0173)	-.0445*** (.0211)	.0657 (.0678)
Farmyard manure (maunds per acre)	.0108* (.0028)	.0140* (.0029)	.0493* (.0031)	.1229* (.0188)
Village Dummies (Noakhali excluded)				
Bogra	-	-3.4675*	-	-
Sylhet	-	1.8542* (.7014)	-	-
Intercept	2.091	3.749	1.175	2.618
Mean of dependent variable	3.27	3.27	2.05	5.46
R^2	.247	.282	.731	.420
R^2 (adjusted)	.232	.263	.721	.381
F statistic	16.134	14.401	71.101	10.968
No. of observations	302	302	164	98

[1] Figures in parentheses are the standard errors of the coefficients.
 * The coefficient is significantly different from zero at the 1% level.
 ** The coefficient is significantly different from zero at the 2.5% level.
 *** The coefficient is significantly different from zero at the 5% level.
 § The coefficient is significantly different from zero at the 10% level.

Table 3.0. Dependent variable: mandays per acre required for fertiliser application under <u>local rice</u> <u>variety: Bangladesh, 1974-75 Boro season.</u>

(Regression coefficients)[1]

Explanatory Variables	Three Villages Aggregate Data		Sylhet Village	Noakhali Village
	Without Village Dummy	With Village Dummy		
Institutional				
Size of operational holding (acres)	-.2045** (.0902)	-.0619 (.0782)	.1501 (.0242)	-.1631 (.3659)
Percentage of area leased-in	-.0056 (.0062)	.0030 (.0054)	-	-.0144 (.0187)
Technical				
Fertiliser dose applied (seers)				
Nitrogen(urea)	.0643* (.0126)	.0374* (.0125)	.0372* (.0060)	.0066 (.0201)
Potash (MP)	-.0451 (.0349)	-.0985* (.0320)	.0124 (.0154)	.0315 (.1164)
Farmyard manure (maunds per acre)	.0086*** (.0042)	-.0103** (.0045)	.0905* (.0053)	.1361* (.0279)
Village Dummies (Noakhali excluded)				
Bogra	-	8.7104* (1.5736)	-	-
Sylhet	-	-2.7493 (.4785)	-	-
Intercept	1.481	3.219	.0623	1.777
Mean of dependent variable	1.6104	1.6104	.447	3.990
R^2	.235	.482	.779	.379
R^2 (adjusted)	.212	.461	.772	.309
F statistic	10.441	22.365	103.210	5.380
No. of observations	176	176	122	50

[1]Figures in parentheses are the standard errors of the coefficients.
*The coefficient is significantly different from zero at the 1% level.
**The coefficient is significantly different from zero at the 2.5% level.
***The coefficient is significantly different from zero at the 5% level.
§The coefficient is significantly different from zero at the 10% level.

Table 31. Dependent variable: mandays per acre required for fertiliser application under HYV rice: Bangladesh 1975 Aman Season

(Regression coefficients)[1]

Data coverage / Explanatory Variables	Three Villages Aggregate Data		Sylhet Village	Noakhali Village
	Without Village Dummy	With Village Dummy		
Institutional				
Size of operational holding(acres)	-.0820 (.1412)	-.0764 (.1543)	-.2176 (.3855)	-.2712 (.2677)
Proportion of area leased-in	-.0163§ (.0098)	-.0109 (.0100)	-.0393§ (.0238)	.0058 (.0103)
Technical				
Fertiliser dose applied (seers per acre)				
Nitrogen(urea)	.0152 § (.0086)	.0089 (.0089)	-.0121 (.0316)	-
Farmyard manure (maunds per acre)	.0201* (.0037)	.0228* (.0039)	.0611* (.0108)	.0839* (.0134)
Village Dummies (Noakhali excluded)				
Bogra	-	-2.0652*	-	-
Sylhet	-	(.7847)	-	-
Intercept	3.995	3.859	3.317	2.205
Mean of dependent variable	3.859	3.859	5.522	3.661
R^2	.140	.194	.501	.309
R^2 (adjusted	.123	.169	.449	.288
F statistic	8.069	7.859	9.557	14.990
No. of observations	203	203	43	104

[1]Figures in parentheses are the standard errors of the coefficients.
*The coefficient is significantly different from zero at the 1% level.
**The coefficient is significantly different from zero at the 2.5% level.
***The coefficient is significantly different from zero at the 5% level.
§The coefficient is significantly different from zero at the 10% level.

Table 32. Dependent variable: Maundays per acre required for
fertiliser application under local rice variety:
Bangladesh, 1975 Aman season.

(Regression coefficients)[1]

Data coverage / Explanatory Variables	Aggregate Data		Bogra Village	Sylhet Village	Noakhali Village
	Without Village Dummy	With Village Dummy			
Institutional					
Size of operational holding (acres)	.1026 (.0785)	.0725 (.0810)	-.0908 (.1071)	.1462 (.1191)	.2186 (.3530)
Percentage of area leased-in.	.0022 (.0053)				.0077 (.0189)
Technical					
Fertiliser dose applied(seers per acre)					
Phosphate (TSP)	.0473* (.0115)	.0227** (.0140)	-	.0902* (.0265)	-.0058 (.0158)
Farmyard manure (maunds per acre)	.0396* (.0031)	.0400* (.0031)	.0227* (.0040)	.0535* (.0046)	.0872** (.0342)

Village Dummies
(Noakhali excluded)

	Without Village Dummy	With Village Dummy	Bogra Village	Sylhet Village	Noakhali Village
Bogra	-	-2.1927* (.6605)	-	-	-
Sylhet	-	-1.2041§ (.6934)	-	-	-
Intercept	.3886	2.1062	.8711	-.2507	2.2185
Mean of dependent variable	2.3383	2.3383	1.5496	3.1139	3.3433
R^2	.3917	.423	.212	.626	.209
R^2 (adjusted)	.3825	.412	.201	.613	.104
F statistic	42.6625	38.7436	18.345	51.226	1.986
No. of observations	270	270	139	96	35

[1]Figures in parentheses are the standard errors of the coefficients.
*The coefficient is significantly different from zero at the 1% level.
**The coefficient is significantly different from zero at the 2.5% level.
***The coefficient is significantly different from zero at the 5% level.
§The coefficient is significantly different from zero at the 10% level.

Table 33. Dependent variable: mandays per acre required for weeding under HYV rice: Bangladesh, 1974-75 Boro season.

(Regression coefficients)[1]

Explanatory Variables	Three Villages Aggregate Data	Bogra Village	Sylhet Village
Institutional			
Size of operational holding (acres)	.2182 (.3543)	-1.2384§ (.8291)	.5237 (.4809)
Percentage of area leased-in.	-	-.1333§ (.0755)	.0305 (.0311)
Technical			
Row planting practised		-4.6206 (4.5723)	-
Number of times weeding practised	1.648§ (1.698)	1.0694 (2.0048)	4.4088* (1.8379)
Fertiliser dose applied (seers per acre)			
Nitrogen (urea)	-	-	-
Phosphate (TSP)			
Village Dummies (Noakhali excluded)			
Bogra			
Sylhet			
Intercept			
Mean of dependent variable			
R^2			
R^2 (adjusted)			
F statistic			
No. of observations			

[1]Figures in parentheses are the standard errors of the coefficients.
*The coefficient is significantly different from zero at the 1% level.
**The coefficient is significantly different from zero at the 2.5% level
***The coefficient is significantly different from zero at the 5% level.
§The coefficient is significantly different from zero at the 10% level

Table 34. Dependent variable: mandays per acre required for weeding under <u>local variety rice</u>: Bangladesh, 1974-75 Boro season.

(Regression coefficients)[1]..

Data coverage Explanatory Variables	Three Villages Aggregate Data	Sylhet Village
Institutional		
Size of operational holding (acres).	-.4880§ (.3219)	-.2386 (.3925)
Percentage of area leased-in.	-	.0087 (.0285)
Technical		
Row planting practised (dummy)	-	-3.7330 (2.9136)
Number of times weeding practised	7.9937* (1.5027)	10.8034* (2.0798)
Fertiliser dose applied (seers per acre)		
Nitrogen (urea)	-	.1988** (.0915)
Phosphate (TSP)	-	-
Potash (MP)	-.1775 (.1218)	-.3482 (.2260)
Farmyard manure applied (maunds per acre)	-	-.0873 (.0825)
Village Dummies (Noakhali excluded)		
Bogra	-2.7065 (5.1386)	-
Sylhet	5.2960 (1.6429)	-
Intercept	3.2090	7.428
Mean of dependent variable	13.234	14.142
R^2	.209	.276
R^2 (adjusted)	.183	.224
F statistic	8.034	5.279
No. of observations	158	105

[1]Figures in parentheses are the standard errors of the coefficients.
*The coefficient is significantly different from zero at the 1% level.
**The coefficient is significantly different from zero at the 2.5% level.
***The coefficient is significantly different from zero at the 5% level.
§The coefficient is significantly different from zero at the 10% level.

Table 35. Dependent variable: Mandays per acre required for weeding under HYV rice: Bangladesh, 1975 Aman season

(Regression coefficients)[1]

Data coverage	Three Villages Aggregate Data		Bogra Village	Sylhet Village
Explanatory Variables	Village Dummy	Village Dummy		
Institutional				
Size of operational holding (acres)	-.2166 (.3320)	-.1457 (.3396)	-.9257 (.6812)	.7771 (.7702)
Percentage of area leased-in	-	.0668 (.0222)	-.0278 (.0498)	.0258 (.0462)
Technical				
Row-planting practised (dummy)	4.0070* (1.9568)	2.5823 (2.2548)	1.7093 (3.2631	-
No. of times weeding practised	3.1735* (.9361)	2.9317* (.9599)	5.8230* (2.1024)	5.0957* (2.1021)
Fertiliser dose applied (seers per acre)				
Nitrogen (urea)	.0147 (.0203)	.0213 (.0209)	-	-.0881 (.1252)
Phosphate (TSP)	.0161 (.0259)	.0054 (.0276)	-.0839 (.1179)	.0478 (.1430)
Potash (MP)	.0440 (.0458)	.0580 (.0472)	.2806*** (.1495)	.3425 (.3280)
Farmyard manure applied (maunds per acre)	-	-	-.0172 (.0138)	-
Village Dummies (Noakhali excluded)				
Bogra	-	-2.4831 (1.9267)	-	-
Sylhet	-	-	-	-
Intercept	4.0498	6.0108	4.4957	.8751
Mean of dependent variable	13.292	13.292	10.580	14.467
R^2	.113	.121	.278	.283
R^2 (adjusted)	.086	.084	.170	.153
F statistic	4.093	3.272	2.581	2.170
No. of observations	199	199	55	40

[1]Figures in parentheses are the standard errors of the coefficients.

*The coefficient is significantly different from zero at the 1% level.

**The coefficient is significantly different from zero at the 2.5% level.

***The coefficient is significantly different from zero at the 5% level.

§The coefficient is significantly different from zero at the 10% level.

Table 36. Dependent variable: .Mandays per acre
required for weeding under <u>local variety</u>
<u>rice crop</u>: Bangladesh, 1975 Aman

(Regression coefficients)[1]

Data coverage / Explanatory Variables	Three Villages Aggregate Data		Bogra Village	Sulhet Village
	Without Village Dummy	With Village Dummy		
Institutional				
Size of operational holding (acres)	-.0906 (.1674)	-.2949[*][**] (.1720)	-.1722 (.1315)	-.4447§ (.2758)
Percentage of area leased-in	-	.0010 (.0113)	-.0125[***] (.0073)	-
Technical				
Row-planting practiced	3.0181[*] (.8375)	.9906 (.8753)	.3229 (.4886)	.8094 (2.2338)
No. of times weeding practised	6.9821[*] (.5634)	5.9110[*] (.5860)	2.7984[*] (.3960)	10.7697[*] (1.1538)
Fertiliser dose applied (seers per acre)				
Potash (MP)	.1248[*] (.0463)	.1729[*] (.0444)	.1603[*] (.0283)	.0495 (.1109)
Village Dummies (Noakhali excluded)				
Bogra	-	-4.4753[*] (1.2279)	-	-
Sylhet	-	.4574 (1.1875)	-	-
Intercept	-1.238	3.309	.643	1.305
Mean of dependent variable	5.416	5.416	1.800	8.314
R^2	.453	.522	.453	.582
R^2 (adjusted)	.445	.507	.429	.560
F statistic	47.609	35.401	18.729	26.485
No. of observations	235	235	119	81

[1]Figures in parentheses are the standard errors of the coefficients.
*The coefficient is significantly different from zero at the 1% level
**The coefficient is significantly different from zero at the 2.5% level
***The coefficient is significantly different from zero at the 5% level
§The coefficient is significantly different from zero at the 10% level

Table 37. Dependent variable: Mandays applied per acre for irrigation under <u>HYV rice crop</u>, Bangladesh 1974-75 Boro season.

(Regression coefficients)[1]

Data coverage Explanatory Variables	Three Villages <u>Aggregate Data</u> Without Village Dummy	With Village Dummy
Institutional		
Size of operational holding (acres)	-2.6142* (0.8486)	-2.8082* (0.8880)
Percentage of area leased-in	-0.1051*** (0.0581)	-0.1097*** (0.0593)
Technical		
Source of irrigation water (percentage of irrigated area under:)		
Power pump	0.0181 (0.0446)	0.0680 (0.0531)
Powered tubewell	0.2942* (0.0524)	0.1299§ (0.0759)
Hand tubewell	1.9001* (0.0812)	1.6163* (0.1250)
Indigenous	0.0712* (0.0280)	0.0807* (0.0283)
Village Dummies (Noakhali excluded)		
Bogra	--	35.7300* (12.3536)
Sylhet	--	2.9055 (5.5900)
Intercept	10.3756	4.4275
Mean of dependent variable	20.3432	20.3432
R^2	0.7007	0.7104
R^2 (adjusted)	0.6942	0.7018
F statistic	106.5432	83.0908
No. of observations	280	280

[1] Figures in parentheses are the standard errors of the coefficients

*The coefficient is significantly different from zero at the 1% level

**The coefficient is significantly different from zero at the 2.5% level

***The coefficient is significantly different from zero at the 5% level

§The coefficient is significantly different from zero at the 10% level

Table 39. Dependent variable: Mandays applied per acre for irrigation under <u>local rice variety</u>, Bangladesh, 1974-75 Boro season.

(Regression coefficients)[1]

Data coverage / Explanatory Variables	Three Villages Aggregate Data	
	Without Village Dummy	With Village Dummy
Institutional		
Size of operational holding (acres)	.0285 (.2307)	-.2290 (.2442)
Percentage of area leased-in	.0193 (.0174)	.0094 (.0174)
Technical		
Source of irrigation water (percentage of irrigated area under:)		
Power-pump	-.0258*** (.0119)	.0040 (.0159)
Deep tubewell	.0662 (.0652)	.1103 § (.0658)
Shallow tubewell	.5778* (.0464)	.6284* (.0489)
Village Dummies (Noakhali excluded)		
Bogra	-	-
Sylhet	-	4.7282* (1.7137)
Intercept	3.8376	-.1487
Mean of dependent variable	3.8207	3.8207
R^2	.544	.566
R^2 (adjusted)	.528	.548
F statistic	35.268	31.972
No. of observations	154	154

[1]Figures in parentheses are the standard errors of the coefficients
*The coefficient is significantly different from zero at the 1% level.
**The coefficient is significantly different from zero at the 2.5% level
***The coefficient is significantly different from zero at the 5% level
§The coefficient is significantly different from zero at the 10% level

Table 39. Dependent variable: Hired mandays per acre of <u>rice cropped area</u>: Bangladesh, 1975 Aman season.

(Regression coefficients)[1]

Data coverage / Explanatory Variables	Three Villages Aggregate Data	Bogra Village	Sylhet Village	Noakhali Village
Institutional				
Size of operational holding (acres)	4.5869* (.5210)	4.9152* (.9860)	3.9174* (.6529)	3.9171** (1.7716)
Percentage of area leased-in.	-.0721* (.0251)	-.0312 (.0512)	-.0776* (.0305)	-.1930* (.0570)
No. of adult agricultural workers in the household.	-1.9112* (.6522)	-2.6904*** (1.324)	-1.1429 (.8777)	-2.1711§ (1.251)
Technical				
Percentage of rice area under HYV	.1581* (.0308)	.3642* (.0876)	.1784* (.0346)	-.0830 (.0590)
Mechanisation level in preparatory tillage (dummy)	.6096 (3.3985)	-	4.2207 (9.1598)	2.2099 (3.2619)
Village Dummies (Noakhali excluded)				
Bogra	6.7073*** (3.4463)	-	-	-
Sylhet	7.9891** (3.3476)	-	-	-
Intercept	13.5501	18.616	4.9022	36.2213
Mean of dependent variable	18.157	21.247	11.517	26.451
R^2	.296	.262	.322	.180
R^2 (adjusted)	.285	.242	.305	.140
F statistic	27.075	12.809	18.691	4.447
No. of observations	459	149	203	107

[1]Figures in parentheses are the standard errors of the coefficients.
*The coefficient is significantly different from zero at the 1% level.
**The coefficient is significantly different from zero at the 2.5% level.
***The coefficient is significantly different from zero at the 5% level.
§The coefficient is significantly different from zero at the 10% level.

Table 40 . Dependent variable: Hired proportion of total farm labour days applied per acre in rice cultivation: Bangladesh 1975 Aman season

(Regression coefficients)[1]

Data coverage / Explanatory Variables	Three Villages Aggregate Data		Bogra Village	Sylhet Village	Noakhali Village
	Without Village Dummy	With Village Dummy			
Institutional					
Size of operational holding (acres)	5.2106* (.8119)	6.4562* (.8618)	6.6272* (1.2573)	5.9914* (1.4886)	8.5745* (2.9292)
Percentage of area leased-in.	-.2020* (.0455)	-.1789* (.0454)	-.1158§ (.0661)	-.2174* (.0822)	-.2740* (.1027)
No. of adult agricultural workers in the household	-6.2120* (1.0878)	-6.4904* (1.0748)	-6.2182* (1.6852)	-6.3087* (2.0482)	-7.2917* (2.0883)
Technical					
Percentage of rice area under HYV.	.1089* (.0389)	.0590 (.0514)	.0755 (.1107)	.0367 (.0736)	.1339 (.1160)
Mechanisation level in preparatory tillage (dummy)	8.3809§ (4.9754)	3.4250 (5.2638)	-	-3.4372 (21.4907)	3.1975 (5.3796)
Village Dummies (Noakhali excluded)					
Bogra	-	-6.7355 (5.6285)	-	-	-
Sylhet	-	-17.6173 (5.4786)	-	-	-
Intercept	59.3933	67.9616	57.9635	52.9107	62.3135
Mean of dependent variable	51.2637	51.2637	46.7190	45.9044	63.299
R²	.2298	.261	.237	.208	.208
R² (adjusted)	.2189	.246	.215	.171	.168
F statistic	21.0640	17.709	10.589	5.662	5.155
No. of observations	359	359	141	114	104

[1]Figures in parentheses are the standard errors of the coefficients.
*The coefficient is significantly different from zero at the 1% level.
**The coefficient is significantly different from zero at the 2.5% level.
***The coefficient is significantly different from zero at the 5% level.
§The coefficient is significantly different from zero at the 10% level.

Table 41. Dependent variable: Hired mandays per acre of
rice cropped area: Bangladesh, 1974-75 Boro
season.

(Regression coefficients)[1]

Data coverage Explanatory Variables	Three Villages Aggregate Data		Sylhet Village	Noakhali Village
	Without Village Dummy	With Village Dummy		
Institutional				
Size of operational holding (acres).	6.6440* (.7105)	6.1747* (.7484)	8.5519* (1.0636)	5.5822* (2.1125)
Percentage of area leased-in.	-.1286* (.0350)	-.1338* (.0358)	-.1550* (.0497)	-.1772* (.0683)
No. of adult agricultural workers in the household.	-2.9302* (.9242)	-2.5736* (.9322)	-5.3602* (1.4285)	-2.6726§ (1.5145)
Technical				
Percentage of rice area under HYV	.2327* (.0280)	.2111* (.0306)	.0356 (.0465)	.1118** (.0548)
Mechanisation level in preparatory tillage (dummy).	-	.8274 (4.8499)	-4.1955 (14.8493)	2.2447 (3.9416)
Village Dummies (Noakhali excluded)				
Bogra	-	-3.9240 (4.0377)	-	-
Sylhet	-	3.9701 (3.7582)	-	-
Intercept	14.7326	15.2563	32.2447	23.4958
Mean of dependent variable	27.277	27.277	35.1532	29.752
R^2	.317	.328	.315	.165
R^2 (adjusted)	.311	.317	.298	.124
F statistic	52.688	31.392	18.129	3.9949
No. of observations	459	459	203	107

[1]Figures in parentheses are the standard errors of the coefficients.
*The coefficient is significantly different from zero at the 1% level.
**The coefficient is significantly different from zero at the 2.5% level.
***The coefficient is significantly different from zero at the 5% level.
§The coefficient is significantly different from zero at the 10% level.

Table 42. Dependent variable: Hired proportion of total
farm labour days applied per acre in <u>rice
cultivation</u>: Bangladesh 1974-75 Boro season

(Regression coefficients)[1]

Data coverage / Explanatory Variables	Three Villages Aggregate Data Without Village Dummy	With Village Dummy	Bogra Village	Sylhet Village	Noakhali Village
Institutional					
Size of operational holding (acres)	5.8096* (.7821)	6.4477* (.8137)	5.2155* (1.8587)	6.9971* (1.0615)	9.5723* (2.8628)
Percentage of area leased-in.	-.2133* (.0419)	-.1881* (0.425)	-.0458 (.1767)	-.2100* (.0490)	-.1794*** (.0917)
No. of adult agricultural workers in the household	-5.4778* (1.1132)	-5.6982* (1.1090)	-3.6072 (4.3417)	-6.4531* (1.4427)	-6.3188* (2.0572)
Technical					
Percentage of rice area under HYV	.1449* (.0396)	.1268* (.0417)	.0515 (.3165)	.1259* (.0475)	.1542*** (.0812)
Mechanisation level in preparatory tillage (dummy)	11.2538* (4.3216)	5.0610 (4.9205)	-	-	4.2090 (5.3118)
Village Dummies (Noakhali excluded)					
Bogra	-	-6.7980 (5.3279)§	-	-	-
Sylhet	-	-10.5536* (3.8719)	-	-	-
Intercept	48.4696	56.0726	49.8989	47.1058	52.2951
Mean of dependent variable	51.711	51.711	56.223	47.314	58.197
R^2	.250	.266	.203	.293	.203
R^2 (adjusted)	.239	.251	.115	.278	.162
F statistic	22.119	17.125	2.297	19.554	4.936
No. of observations	338	338	41	194	103

[1]Figures in parentheses are the standard errors of the coefficients.
*The coefficient is significantly different from zero at the 1% level.
**The coefficient is significantly different from zero at the 2.5% level.
***The coefficient is significantly different from zero at the 5% level.
§The coefficient is significantly different from zero at the 10% level.

Chapter 5
=========

Access to Factor Markets,
=========================
Knowledge on HYV and
====================
Farmers' Organisations
======================

Introduction

         Existence of a high degree of factor market imperfections
in the rural areas of most developing countries is, by now,
fairly well known [see 37,13]. The main feature and manifestation
of this imperfection is that access to factors of production is
much easier for some groups than others. In the absence of
technical change, existence of such a phenomenon would affect
the allocation of resources in agriculture, the methods of production adopted by the farmer, and the distribution of rural
income. Following technological change, when there is a greater
need for credit, improved seeds, fertilisers, irrigation water,
extension services (knowledge of the new technology), participation
in farmers' organisations, etc., question of farmers' access to
these factors would be of crucial significance.
         Therefore we begin this chapter by investigating the relative
importance of the communication channels of information on HYV
and the credibility of the different sources to the farmer. The
relative access to indigeneous and formal channels of communications
by different farm size groups is also investigated. The problem
of access to factor market studied in relation to the HYV technology, covers two important factors - capital and chemical fertilizers. The issue of access to the product market is examined
in relation to the prevailing agrarian structure and change in
technology. Finally, the important question of access to farmers'
organisations is taken up and the background of the participating
farmers is studied in terms of farm size, tenancy and the level
of farmer's education.

Before we undertake the study of the above issues with the help of bivariate and multivariate analyses of the cross-sectional survey data, a brief description of the theoretical and analytical framework is discussed below.

Access to Knowledge on HYV

It has been pointed out in Chapter 2 that the cultivation of HYV brings with it new inputs whose rate and timing of application and various combinations over the cropping season are highly complex and calls for the use of improved cultural practices. Thus the farmers are generally uncertain about their returns from land in such a situation. While the risk and uncertainty attached to HYV cultivation exists for farmers of all sizes, the degree of uncertainty is higher for the smaller farms. This is partly because the effectiveness of the communication channels for disseminating knowledge and information on the new technology is limited. Moreover, access to knowledge of the use of the new technology is imperfectly distributed between the small and large farmers.

Therefore an assessment needs to be made of the relative effectiveness of alternative communication channels such as (a) indigenous communication system (neighbours, friends, relatives, etc); (b) change agents (agricultural extension agencies, model farmers from farmers' organisations, etc.) and (c) mass media (farm radio programme).

The process of adoption of innovations by cultivators is widely accepted to consist of five stages and these are awareness, interest, evaluation, trial and adoption. In this study we shall analyse the two most important (initial and final) stages of the process of HYV adoption. In order to evaluate the relative effectiveness of the various channels of communication at the awareness stage, the question asked was: "From whom or what sources did you first come to know about HYV rice?"

Furthermore, knowledge on innovations may reach the farmers
through numerous channels but media credibility (faith or belief
cultivators placed on any one source of information as compared
to others) would assume considerable importance at the adoption
stage. Hence, we investigated the relative credibility of the
various sources of information on HYV by asking the question
"From whom or what sources did the information on HYV fully
convince you to try it?"

In one part (Tamil Nadu) of India the most commonly
reported source of knowledge quoted by both HYV adopters and
nonadopters was "neighbours and other farmers" and the
agricultural extension workers was quoted as the second
important source [23]. In another part (Western Rajasthan) of
India, interpersonal communication along informal channels
was found to be the most important [18]. Evidence from a third
region (Andhra Pradesh) of India reveals that informal personal
sources and change agents were the two important sources [56].
By far the most widely covered survey of HYV growers in India
revealed that the government extension agencies were reported
as information source by over one half of the farmers; a
quarter of them reported other cultivators as their source [35].[1]
Similarly in one province (South Cotabato) of the Philippines,
80% of the farmers' first information on HYV was from other
farmers; while in two other provinces (Caramines Sur and Iloilo)
government extension agencies were the first source of information for over half the farmers there and "friends and relatives"
were reported as the first source of information on HYV by about
30 to 40% of the cultivators [48].

Therefore it is expected that the applied educative services
of the government (government extension agencies, mass media,
model farmers of cooperative societies trained under the auspices
of the government's Integrated Rural Development Programme)
are relatively more effective in channelling information and

---

[1] Farms covered under this survey were drawn from the
States of Andhra Pradesh, Assam, Bihar, Jammu and Kashmire,
Kerala Namilnadu, Maharashtra, Mysore, Orissa, Punjab, U.P.
and West Bengal.

knowledge on the new technology. The rationale behind such a
hypothesis is (i) skills and knowledge thus acquired are readily
incorporated into farmers' production system; (ii) no problems
of drop-outs or training people for jobs they will not do;
(iii) the time lag between learning and practice is likely to
be days and not years, and (iv) a practising farmer is better
motivated. [16].

Access to Factor Markets

The cultivation of HYVs requires the purchase of modern
inputs (seed, fertiliser, irrigation water, etc.) from the
market and in order to raise the necessary cash, there is
often the need to borrow money. Traditionally, smaller farmers,
because of lower family incomes, have always been more dependent
on credit than larger farmers, the latter often functioning as
money lenders advancing money at high interest rates. The
smaller farmer has fewer assets to offer as collateral. Moreover, the larger farmer's ability to repay loans particularly to
institutional sources further enhances his credit worthiness.
This reduces the small farmer's credit worthiness. All these
result in larger farmers' easier access to cheaper sources of
credit. These conditions also lead to the small farmers paying
a higher price for borrowed funds.

Chemical fertiliser is one of the most important factors
of production associated with the HYV technology. In Bangladesh
it is a subsidised input distributed through a network of
government appointed dealers at the village level and through
the village cooperatives. In 1977-78 the subsidy as percentage
of cost was 48% for urea (nitrogen) fertiliser, 67% for
phosphate (TSP) and 60% for potash (MP) [34].[1] The larger
farmers in view of their high social and economic status (and
often political power) in the villages are the first to receive
the supplies at the officially determined subsidised prices.

---

[1]These subsidies are based on the difference between price
and actual cost and not on the evaluation of social cost.

Since the small farmers are treated residually, their access
to officially appointed retail sources are limited and they are
compelled to meet their needs through purchases made at higher
prices from the black market.

Marketable Surplus

Generation of marketed surplus is considered as some kind
of a measure of efficiency. The introduction of the new technology has led to an increase in the proportion of total rice
harvested sold in the market [26,43]. Traditionally, the large
farmers are also the surplus producers. Moreover, the smaller
farmer is likely to get a lower price of his rice produce sold
compared to his large neighbours, due to his lack of storage
facilities and inability to hoard the grain until the price is
favourable.

Access to Farmers' Organisations

The expansion of the Comilla-type cooperatives under the
Bangladesh Government's Integrated Rural Development Programme
(IRDP) as the organisational framework was believed to be an
appropriate institution for combining growth with equality
during the transition to the cultivation of HYVs. The programme
has expanded quite rapidly covering 46% of the administrative
thanas.[1] The basis of the programme was chanelling of modern
inputs (e.g. low-lift irrigation pumps, tube wells, improved

seeds, fertiliser, pesticide, knowledge and extension services)
which were very heavily subsidised. For example, in addition to
the subsidies on fertiliser indicated earlier, in 1973-74, the
rate of subsidy for low-lift pump irrigation was Taka 107 per
acre of irrigated area and it was 68% of total cost; for deep
tubewell the rate of subsidy was Taka 166 per acre of irrigation
and 77% of total cost [43]. Survey of evidence drawn from a

----

[1]A thana in Bangladesh is a lower level **administrative unit**
with an average area of 125 square mil**es** and an average population
of about 180 thousand. There are 411 thanas in all.

number of areas of Bangladesh suggests a higher participation
rate for large farmers in the membership and management of
these organisations [3, 43]. Several reasons have been put
forward for such unequal participation. It is believed that
the modest conditions of membership were too difficult for the
poor peasants to fulfill. It is also probably that the poor
peasants recognised that the structure was such that the
benefits would at best be proportional to initial landholding.
Another reason given is the poor farmer's fear of the unequal
partnership with the large farmer. Finally, the programme had
virtually no place to offer to the landless labourers.

Similarly, the tenants participation rate in these organisations are likely to be lower. However, farmers with higher
levels of education are likely to have a bigger participation
rate due to the innovative effect of education on farmers (as
discussed in Chapter 2).

Hypotheses

The hypotheses that are being tested can be summarised as
follows. Farmers use in some proportion both indigeneous
channels of communication and change agents as sources of information on HYV at both the awareness and adoption stages. Mass
media (farm radio programme) is not expected to serve as a
significant source of first information (awareness stage) and
would be even less significant at the adoption stage. Degree
of reliance on indigeneous channels for information on HYV
is expected to be negatively associated with farm size. Conversely, the degree of reliance on official change agents
(government extension services) is expected to be positively
associated with farm size.

In the capital market it is hypothesised that the amount
borrowed per acre is <u>negatively</u> associated with farm size and
with tenancy. Credit worthiness is expected to be higher among
larger farmers because (i) income is expected to be <u>positively</u>
related to farm size, (ii) number of farm assets owned is
expected to <u>positively</u> related to farm size, (iii) amount of
loan repaid is <u>positively</u> associated with farm size and (iv)

the proportion of the total repayment made going to institutional sources is <u>positively</u> associated with farm size. The
amount of loan obtained per acre from institutional sources is
expected to be <u>positively</u> associated with farm size and conversely,
the amount obtained per acre from noninstitutional sources is
expected to be <u>negatively</u> associated with farm size. Finally,
the cost of borrowing (rate of interest) is expected to be
<u>negatively</u> associated with farm size.

In the fertiliser market, the percentage of fertiliser
purchase from the black market (at higher cost) is expected
to be <u>negatively</u> related to farm size.

In the product market the percentage rice output sold in
the market is expected to be <u>positively</u> associated with farm
size and HYV adoption; it is expected to be <u>negatively</u> associated
with tenancy. The price received for rice output sold is expected to
be <u>positively</u> associated with farm size and <u>negatively</u> with tenancy.

Finally, participation in farmers' organisations is expected
to be <u>positively</u> associated with farm size and level of education;
it is expected to be <u>negatively</u> associated with tenancy. Amount
borrowed per acre from institutional sources is expected to be
<u>positively</u> associated with membership in farmers' organisations;
conversely, the amount borrowed from noninstitutional sources
is expected to be negatively associated with membership in
farmers' organisations.

Empirical Verification

In the following section we undertake the empirical
verification of the above hypotheses by analysing farm level
cross-sectional variations in access to knowledge on HYV with
the help of bivariate tables. Variations in the access to the
capital, fertiliser and product (during Aman and Boro seasons)
markets are analysed primarily by means of multiple regression
analysis. Inter-farm variations in the membership in farmers'
organisations is analysed with t he help of a special type of
multivariate analyses known as logit analysis which is applied
when the dependent variable is binary (see text and appendix

for elaborations). In virtually all the investigations, regional variations are also taken into account. We begin with the issue of access to knowledge on HYV.

Access to Knowledge on HYV

We discuss the results in two parts. First, the question of the choice and credibility of communication channels is analysed, followed by the analysis of the relationship between farm size and the choice and credibility of the communication media used.

Choice and Credibility of Communication Channels: Nearly one half of the cultivators reported government extension agencies as the "most important" source of first information on HYVs (Table 43). Neighbour ranked next with over 40% of the cultivators reporting them as a "most important" source of first information. However, nearly 60% of the cultivators reported both these channels of information either as "most important" or as "important" for first information on HYV. In addition to the importance of these two sources of first information on HYV, the credibility of these two sources to the cultivators was high. The farmers placed the most reliance on government extension agencies in making their decision to adopt HYV (50% of the cultivators reported it as "most important"; nearly 63% of the cultivators reported it either as "most important" or as "important"). Farmers' reliance on indigeneous sources in their decision to adopt HYV ranked second (nearly 35% of the cultivators reported "neighbours" as the "most important" source during the adoption stage; over 55% of the farmers reported this source either as "most important" or as "important").

Mass media (farm radio programme) performed very poorly as a first source of information (90% of the cultivators reported this source as "unimportant") and it had much less success in convincing the farmers to adopt HYV (its extremely low credibility level is revealed by the fact that nearly 97% of the cultivators reported this source as "unimportant" at the adoption stage).

This is little surprising; only 7% of the 625 households surveyed
owned a radio set. However, attendance to radio programmes **was**
much higher. Two-thirds of the respondents listened to radio
programmes, a fifth of them regularly. A fifth of the respondents
never listen to the radio.

Farm size, choice and credibility of media: As expected, a
significant negative association is observed between the proportion
of cultivators using indigeneous communication channels ("neighbours"
as a first source of information on HYV and farm size. This
negative association is observed when this source is described
exclusively as "most important" as well as when it is described
in terms of "most important" and "important". (Table 44). A
significant positive association is observed between the proportion of farmers using formal channels of communication of
information on HYV and farm size. This positive relationship is
observed when farmers reported this source individually as "most
important" and when they reported it together as "most important"
and "important". Similar associations between farm size and proportion of farmers relying on these two sources is observed at
the adoption stage implying that smaller farmers attach greater
faith to information obtained through indigeneous channels while
the larger farmers attach higher credibility to information obtained
from official change agents.

Access to Factor Markets

     We begin our study of the factor markets by first looking
at the capital market.

Capital Market: During May 1974 - April 1976, farmers borrowed
on an average amount of Taka 1327 per acre of which 88% was
obtained from noninstitutional sources (Table 45). Over 80% of the
282 borrowers obtained their loans from noninstitutional sources.
While the average cost (annual rate of interest) of borrowing
from institutional sources was about 9% per annum, it was 23%
for amounts borrowed from noninstitutional sources. Variations
in the cost of borrowing were much wider for noninstitutional
sources (coefficient of variation of the annual rate of interest
was .66 for institutional sources as opposed to 2.27 for non-

instituional sources).

As expected, total amount borrowed (per acre) is <u>negatively</u>
associated with farm size; no significant association with
tenancy or membership in farmers' organisations is observed
(Table 45). What is more interesting is that while no significant
relationship between amount of borrowing (per acre) from institutional sources and farm size is observed, amount borrowed (per
acre) from noninstitutional sources is <u>negatively</u> related to farm
size. As expected, amount borrowed (per acre) from institutional
sources was significantly <u>higher</u> for members of farmers' organisations; <u>conversely</u>, the amount borrowed (per acre) from noninstitutional sources was significantly <u>higher</u> for nonmembers.
Amount of borrowing (per acre) had no significant relationship to
the household size. Significant regional difference in borrowing
is observed with households in the Bogra and Sylhet villages
borrowing significantly less than those of Noakhali.

From the multiple regression equation below, it is clear
that the cost of borrowing is <u>negatively</u> related to farm size.[1]

$$i = 14.56 - 6.15A* + 29.35R_2^* - 3.29R_1$$
$$\quad\quad\quad (2.14) \quad\quad (8.17) \quad\quad (16.29)$$

$$N = 191 \quad\quad\quad R^2 = .091$$

where i is the annual rate of interest in percent, A is the size
of the operational holding in acres, R_2 is the regional dummy for
the Sylhet village (taking a value of 1 if the borrower belongs
to this village, 0 otherwise) and R_1 is the regional dummy for the
Bogra village (figures in parentheses are the standard errors
of the coefficient and N is the number of observations).

Following are the major features of the measures of credit-
worthiness examined:

(a) As expected, gross income per household is <u>positively</u>

---

[1]This relationship is tested in terms of costs of noninstitutional loans since more than 80% of the 282 borrowers obtained
88% of the amount borrowed from noninstitutional sources.

*The coefficient is significantly different from zero at the
1% level.

related to farm size (see regression equation below).[1]

$$y = 59.35 + 18.52A^* - 0.15T + .50F$$
$$\quad\quad\quad (1.99) \quad\quad (0.10) \quad (8.03)$$

$$\quad\quad - 11.36H - 31.76R_1^* - 28.31R_2$$
$$\quad\quad\quad (9.36) \quad\; (10.82) \quad\; (10.79)$$

$$N = 459 \quad\quad\quad\quad R^2 = .20$$

where y is gross income per household during the 1975-76 agricultural year in Takas; as before, A, R_1 and R_2 respectively denote size of operational holding in acres and regional dummies for the Bogra and Sylhet villages; T is the percentage of area leased-in; F is a dummy for participation in farmers' organisations (it assumes a value of 1 if the farmer is a member, 0 otherwise) and H is the HYV adoption dummy (it assumes a value of 1 if the farmer adopts HYV, 0 otherwise).

(b) As expected, a significant _positive_ correlation was observed between farm size on the one hand and on the other, ownership of some selected farm assets like livestock (bullocks, cows and calves), farm equipment (plough, hoes and weeder), irrigation pump, fishing net, water transport (boat) and radio sets (Table 46).

(c) As expected, the absolute amount of loan repaid is _positively_ related to farm size (Table 47). Furthermore, credit worthiness, particularly with institutional sources improves further for the larger farmers as the repayment of institutional loans as percentage of total repayment is _positively_ associated with farm size and membership in farmers' organisations and it is _negatively_ associated with tenancy (Table 47). Significant regional differences are also observed.

Fertiliser Market: It appears that about 41% of the urea fertiliser purchasers obtained their supplies from the black market by paying

---

[1] The average gross income per household was Taka 5,836.- for the agricultural year 1975-76. It is also clear from this equation that after controlling for tenancy, membership in farmers organisations and adoption of HYV technology, the significant negative relationship between income and farm size remains unaltered. It is also seen that no significant association is observed between farm size and tenancy or participation in farmers' organisations or adoption of HYV.

*The coefficient is significantly different from zero at the 1% level.

two and a half times the official price (Table 48). About a third of
the other two fertiliser (potash and phosphate) purchasers obtained their supplies from the black market at over one and a
half times the price at the officially appointed dealer's shop.
On the average, the black market provided 63% of the total
quantity of fertiliser purchased by the farmers. The following
regression equation shows that the smaller farmers are more
dependent on the black market for their supplies of fertilisers.
The proportion of total fertiliser purchased from the black
market is <u>negatively</u> related to farm size.

$$Q = 50.41 - 3.53A^* + 6.48F + 1.41H$$
$$(1.07) \qquad (6.11)$$
$$+ 28.40R_1^*$$
$$(4.73)$$

$$N = 235 \qquad\qquad R^2 = .19$$

where Q is the quantity of fertiliser purchased from the black
market as a percentage of the total fertiliser purchased; A, F,
H and R_1, as before, respectively denote the size of operational
holding in acres, membership in farmers' organisations (dummy),
adoption of HYV (dummy) and regional dummy for Bogra.

Marketable Surplus

Over 8% of the Aman rice output and about 18% of the
Boro rice output was sold in the market (Table 49). As expected,
percentage of rice output marketed during both the Aman and
Boro seasons is <u>positively</u> associated with farm size; it is
<u>negatively</u> associated with tenancy. No significant association
was observed with membership in farmers' organisations. During
the Boro season percentage of output marketed is significantly
higher for the adopters compared to the nonadopters. Contrary
to expectations, the correlation coefficient between percentage
of Boro rice output marketed and farm size was significantly
<u>negative</u> (−.13). However, during the Aman/season this coefficient
was positive (.04) although not significant.

Access to Farmers' Organisations

Neary 50% of the 625 households are members of farmers' organisations. Nearly 60% of the 304 members of the farmers' organisations belong to the Comilla-type village cooperatives (KSS) and other multipurpose cooperative societies. Another 40% of the members of farmers' organisations belong to irrigation groups. By using a special type of multivariate analysis, the probability of an individual cultivator becoming a participant in farmers' organisations is examined. Under this approach, the dependent variable for each observation takes a value of one if the respondent has joined a farmer organisation, while a value of 0 is assigned to those individual cultivators who have not joined. The dependent variable being dichotomous, the appropriate type of multivariate analysis which we are using is logit analysis (a description of this statistical technique is presented in the appendix). The results are presented in Tables 50 and 51. While Table 50 presents results based on a mixture of continuous and binary explanatory variables, results based exclusively on binary explanatory variables are presented in Table 51. The latter table permits us to rank the individual explanatory variables in terms of their relative contributions in explaining the variations of the dependent variable.

As expected, the propensity to participate in farmers' organisations is _positively_ associated with farm size. Both large (5.5 acres or above) and medium (2.5 - 5.5 acres) sized farms are more likely to participate than the small farms. Contrary to our expectation when tenancy is defined as percentage area leased-in, propensity to participate is _negatively_ associated with it. However, when tenancy is used as a binary explanatory variable, no significant difference in the propensity to participate is observed between the owner and tenant farms. It is interesting to note that the owner-cum-tenant farms are more likely to participate than the tenant farms.

Farmers with primary or higher levels of education are more likely to participate compared to those who are illiterate or literate or educated below primary level. When the explanatory binary variables are ranked in terms of their explanatory power, primary education level or above ranks first, followed by owner-cum-tenant farms, literally, medium size farms, less than primary level of education and finally large sized farms.

Significant regional differences in the propensity to participate are observed. The Noakhali farmers are more likely to participate than those in Bogra and Sylhet.

Summary and Conclusions

The high degree of credibility accorded to both indigeneous channels of communication ("neighbours") and government extension agencies speaks of the need for greater utilisation of local channels of communication through local leaders and village level extension agencies. Indigenous channels are particularly important for smaller farms while the official extension agencies appeared to be biased towards disseminating knowledge on HYV technology to the larger farmers.

Reliance on the non-institutional sources of credit, in terms of amount borrowed (88%) and the number of borrowers (80%), is very high and this means the costs of borrowing from the rural capital market is also high (annual rate of interest is 23% for this source as compared to 9% for loans obtained from institutional sources). While the reliance and cost of working capital is high for all cultivators, it is greater for the smaller farms. Rate of interest is <u>negatively</u> associated with farm size and amount of loans (per acre) obtained from institutional sources is also <u>negatively</u> related to farm size. Small farmers are more dependent on borrowings than the large farmers. Total amount borrowed is <u>negatively</u> associated with farm size. The existence of these imperfections in the factor market is not surprising. Credit worthiness is positively related to farm size. For example, gross income is positively related to farm size; ownership of farm assets (which provides ability to offer collateral for loans) is <u>positively</u> related to farm size; and loan repayment records particularly with respect

to institutional sources is better for larger farms, implying
defaults on loans by this group.

Between one-third to two-thirds of the fertiliser purchasers
obtain their supplies from the black market by paying between
one and a half to two and half times the official dealer's price
and the fertiliser purchasers are reliant on the black market for
the bulk (63%) of their supplies. While the cultivators
as a whole are in the grips of the black market, the smaller
farms dependency on this costlier market is greater. Amount
of fertiliser purchased from the black market as a percentage
of total amount purchased is negatively related to farm size.

The proportion of rice output marketed is positively associated with farm size; it is negatively associated with tenancy.
Membership in farmers' organisations does not appear to have any
influence on the marketable surplus. The green revolution technology appear to generate marketable surplus. Seasonal variation
in percentage rice output marketed is observed (during Boro, it
is 20% as compared to 8% during Aman). Contrary to expectations,
a significant negative correlation between price of rice output
sold and farm size is observed during the Boro season.

The access to farmers' organisations appears to be quite
unequal. Larger farmers and medium sized farmers are more likely
to participate in farmers organisations than smaller farmers;
owner-cum-tenants are more likely to participate than pure tenants;
farmers with primary or higher levels of education are more likely
to participate than others with none or lower-level education.

Table 43. Communication Channel Use in Relation to the Awareness
and Adoption Stages of HYV: Bangladesh 1976.

Stage, Media credibility level Source of Information	Awareness Stage (N=455) (% of cultivators)			Adoption Stage (N = 365) (% of HYV Adopters)		
	Most Important	Important	Unimportant	Most Important	Important	Unimportant
Neighbours	40.2	18.7	41.1	34.8	20.5	44.7
Friends/ Relatives	4.6	14.8	80.6	6.9	15.7	77.5
Model Farmer	4.6	8.2	87.2	4.1	7.5	88.4
Irrigation Group Manager	3.5	1.8	94.7	9.7	6.6	83.7
Government Extension Agency	47.6	11.0	41.4	49.3	13.4	37.3
Farm Radio Programme	4.0	5.5	90.5	0.3	3.0	96.7

Table 44. Source of First Information
on HYV and Media Credibility
by Farm Size: Bangladesh 1976.
(% of farms under each category)

Farm Size (acres)	Indigeneous Communication Channel (Neighbours)								Formal Communication Channels (Government extension agency)							
	Awareness Stage				Adoption Stage				Awareness Stage				Adoption Stage			
	No. of Farms	Most Impor-tant	Impor-tant	Un-impor-tant	No. of Farms	Most Impor-tant	Impor-tant	Un-impor-tant	No. of Farms	Most Impor-tant	Impor-tant	Un-impor-tant	No. of Farms	Most Impor-tant	Impor-tant	Un-impor-tant
0.0 - 0.5	90	70.0	13.3	16.7	58	43.1	19.0	37.9	90	30.0	6.7	63.3	57	36.8	12.3	50.9
0.5 - 1.0	96	44.8	18.8	36.5	74	37.8	24.3	37.8	95	41.1	12.6	46.3	74	40.5	12.2	47.3
1.0 - 1.5	71	45.1	22.5	32.4	51	43.1	15.7	41.2	72	41.7	13.9	44.4	52	42.3	5.8	51.9
1.5 - 2.0	60	28.3	18.3	53.3	51	27.5	23.5	49.0	60	50.0	16.7	33.3	51	56.9	13.7	29.4
2.0 - 2.5	29	34.5	20.7	44.8	27	33.3	22.2	44.4	29	48.3	13.8	37.9	27	48.1	18.5	33.3
2.5 - 3.0	25	28.0	20.0	52.0	23	13.0	17.4	69.6	25	52.0	8.0	40.0	23	60.9	8.7	30.4
3.0 - 3.5	31	16.1	19.4	64.5	28	39.3	25.0	35.7	30	76.7	3.3	20.0	28	57.1	17.9	25.0
3.5 - 5.5	31	12.9	25.8	61.3	31	25.8	16.1	58.1	31	77.4	3.2	19.4	31	67.7	16.1	16.1
5.5 and above	22	9.1	13.6	77.3	22	31.8	18.2	50.0	22	72.7	18.2	9.1	22	63.6	27.3	9.1
All Sizes	455	40.2	18.7	41.1	365	34.8	20.5	44.7	454	47.6	11.0	41.4	365	49.3	13.4	37.3
Chi-Square (χ^2)	74.846				16.204				54.979				33.741			
Level of Significance	0.0001				0.439				0.0001				0.006			

Table 45. Dependent Variable: Amount Borrowed
per acre (Takas) Bangladesh May 1974–
April 1976

(Regression Coefficients)[1]

Source of
loan

Explanatory Variables	Total	Institutional	Non-institutional
Size of operational holding (acres)	-352.4* (105.5)	-45.0 (31.5)	-310.1* (104.6)
Percentage of area leased-in	-6.0898 (3.8955)	-1.007 (1.174)	-5.0776 (3.9009)
Membership in farmers' organisations (dummy)	-399.0 (314.0)	231.5* (93.7)	-628.4*** (311.3)
Size of household	70.9 (52.9)	11.984 (15.820)	61.6 (52.6)
Village Dummies (Noakhali excluded)			
Bogra	-1766.1* (469.5)	18.395 (140.741)	-1796.8* (467.6)
Sylhet	-1160.6* (377.5)	-29.898 (113.351)	-1139.7* (376.6)
Intercept	2720.1	73.8	2640.5
Mean of dependent variable	1327.0	160.7	1162.3
R^2	0.1803	0.0608	0.1672
R^2 (adjusted)	0.1575	0.0344	0.1439
F statistic	7.9165	2.3074	7.1618
No. of observations	223	221	221

[1] Figures in parentheses are the standard errors of the coefficients

* The coefficient is significantly different from zero at the 1% level
** The coefficient is significantly different from zero at the 2.5% level
*** The coefficient is significantly different from zero at the 5% level
§ The coefficient is significantly different from zero at the 10% level

Table 46. Ownership of Selected Farm Assets by Farm
Size: Bangladesh 1976

(Number of farm homeholds = 459)

Type of Farm Asset	Value of Correlation coefficient between farm size and number of farm assets owned	F-statistic	Level of Significance of F
Livestock			
Bullocks	+.58	32.71	.0001
Milch cows	+.47	17.57	.0001
Cows	+.31	7.69	.0001
Calves	+.52	24.68	.0001
Farm Equipment			
Plough	+.67	53.64	.0001
Hand-hoe	+.18	2.17	.029
Bullock-hoe	+.12	3.22	.001
Hand-weeder	+.29	6.99	.0001
Irrigation equipment			
Hand tubewell	+.27	6.80	.0001
Fishing equipment			
Fishing net	+.44	16.00	.0001
Transport			
Boat	+.43	18.11	.0001
Mass media			
Radio set	+.28	5.60	.0001

Table 47. Dependent Variables: Loan
Repayment by Amount and Source:
Bangladesh May 1974 - April 1976
(Regression coefficients)[1]

Explanatory Variables	Amount of loan repaid (Takas)	Proportion of total loan repaid to Institutional sources (% of total amount repaid)
Size of operational holding (acres)	135.78* (23.5)	3.0250*** (1.5222)
Percentage of area leased-in	-0.4630 (1.2195)	-0.3808* (0.0986)
Membership in farmer's organisation (dummy)	-110.0 (96.9)	36.4* (7.4)
Village Dummies (Noakhali excluded)		
Bogra	-375.9* (120.1)	48.7* (9.7)
Sylhet	-147.7 (129.7)	26.0* (9.1)
Intercept	326.0	10.4
Mean of dependent variable	317.4	54.9
R^2	0.1040	0.3357
R^2 (adjusted)	0.0942	0.3149
F statistic	10.5210	16.1705
No. of observations	459	166

[1]Figures in parentheses are the standard errors of the
coefficients

*The coefficient is significantly different from zero at the
1% level

**The coefficient is significantly different from zero at the
2.5% level

***The coefficient is significantly different from zero at the
5% level

§The coefficient is significantly different from zero at the
10% level

Table 48. Cost of fertilizer purchased from the official
dealer and from the Black Market: Bogra, Sylhet
and Noakhali Districts of Bangladesh 1974-75.
(Takas per feer)

Type of fertiliser	Source Price	Official dealer		Black Market		Index for Black Market price (official price = 100)
		No. of purchasers	Price	No. of purchasers	Price	
Urea		290	1.20	207	3.20	258
Potash		120	0.85	53	1.35	158
Phosphate		207	1.00	109	1.58	158

Table 49. Dependent Variable:
Percentage of rice output
sold in the market:
Bangladesh 1975 Aman and
1974-75 Boro Seasons

(Regression coefficients)[1]

Explanatory Variables	Aman 1975	Boro 1974-75
Size of operational holding (acres)	0.8377*** (0.4426)	2.2162* (0.5611)
Percentage of area leased-in	-0.0508* (0.0249)	-0.0635*** (0.0331)
Adopter of HYV (dummy)	2.6184 (1.9772)	10.5747* (3.7020)
Membership in farmer organisation (dummy)	-2.2083 (1.8850)	-1.2478 (2.7569)
Village Dummies (Noakhali excluded)		
Bogra	-9.0863* (2.4407)	-5.3369 (3.9274)
Sylhet	-3.8379 (2.8417)	5.3687§ (3.1810)
Intercept	12.4577	3.0864
Mean of dependent variable	8.1048	17.6528
R^2	0.1199	0.1334
R^2 (adjusted)	0.1039	0.1172
F statistic	7.4959	8.2353
No. of observations	337	328

[1]Figures in parentheses are the standard errors of the coefficients

*The coefficient is significantly different from zero at the 1% level

**The coefficient is significantly different from zero at the 2.5% level

***The coefficient is significantly different from zero at the 5% level

§The coefficient is significantly different from zero at the 10% level.

Table 50. Maximum Likelihood Estimation
of Dichotomous Logit Relationship.[1]
Bangladesh 1976

(Dummy Dependent Variable = 1 if the
cultivator is a member of farmer
organisation, 0 otherwise)[2]

Explanatory Variables	Data Coverage Without Village Dummy	With Village Dummy
Institutional		
Size of operational holding (acres)	0.1325*** (2.0468)	0.5146* (5.5460)
Percentage of area leased-in	0.0004 (0.1295)	0.0078** (2.1962)
Educational Dummies (Above primary level excluded)		
Illiterate	-1.9219* (-6.5807)	-1.9472* (-5.7367)
Literate	-1.0154* (-2.7051)	-1.4804* (-3.3645)
Up to Primary level	-0.5328 (-1.4480)	-0.8923*** (-2.0811)
Village Dummies (Noakhali excluded)		
Bogra	--	-2.5178* (-5.8871)
Sylhet	--	-4.1093* (-8.8375)
Intercept	1.1440* (3.9570)	3.2075* (6.6968)
R^2	0.1573	0.3825
No. of observations	456	456

[1] Description of this statistical technique is presented in the
appendix

[2] Figures in parentheses are the t-statistic of the coefficients

* The coefficient is significantly different from zero at the
1% level

** The coefficient is significantly different from zero at the
2.5% level

*** The coefficient is significantly different from zero at the
5% level

§ The coefficient is significantly different from zero at the
10% level

Table 51. Maximum Likelihood Estimation of
Dichotomous Logit Relationship[1]:
Bangladesh 1976
(Dummy Dependent Variable = 1 if the
cultivator is a member of farmer
organisation, 0 otherwise; and
Explanatory Binary Variables)
(Standardized value of coefficients)[2]

Explanatory Binary Variables	Data Coverage Without Village Dummy	With Village Dummy
Institutional		
Farm Size Dummies (Small farms, i.e. less than 2.5 acres excluded)		
Large Farms (over 5.5 acres)	0.0056 (0.4922)	0.0291* (2.6087)
Medium Farms (2.5 - 5.5 acres)	0.0013 (0.0553)	0.0701* (2.9231)
Tenancy Dummies (Pure tenants excluded)		
Owner farms	0.2017* (2.5290)	0.0811 (0.9185)
Owner-cum-tenants	0.1554* (2.7730)	0.1187*** (1.9565)
Educational Dummies (Above primary level excluded)		
Illiterate	−0.4422* (−6.8010)	−0.4100* (−6.1695)
Literate	−0.0589* (−2.9442)	−0.0725* (−3.4954)
Less than Primary Level	−0.0377§ (−1.5648)	−0.0549** (−2.2158)
Village Dummies (Naokhali excluded)		
Bogra	--	−0.3145* (−5.6080)
Sylhet	--	−0.6385* (−8.1211)
R^2	0.1663	0.3482
No. of Observations	456	456

[1] Description of this statistical technique is presented in the appendix
[2] Figures in parentheses are the t-statistic of the coefficients
*The coefficient is significantly different from zero at the 1% level
**The coefficient is significantly different from zero at the 2.5% level
***The coefficient is significantly different from zero at the 5% level
§The coefficient is significantly different from zero at the 10% level

Chapter 6

Conclusions and Policy Implications

Summary of Findings

One major finding of this study is that while one observes an inverse relationship between overall land productivity and farm size in a traditional agricultural setting, the introduction of the green revolution technology has either weakened or eliminated this inverse relationship.[1] Tenancy appears to significantly depress land productivity on the traditional farms; the introduction of the green revolution technology tends to strengthen this inverse relationship between productivity and tenancy.

The second major finding of the study is that while an inverse relationship between farm size and overall babour intensity per acre is virtually nonexistent under traditional agriculture, the introduction of the green revolution technology leads to the emergance or strengtheni of the inverse relationship between farm size and per acre labour-use. The depressing effect of tenancy on labour use is significant under the new technology.

The higher risk and uncertainty associated with the cultivation of HYV crop during Aman compared to Boro are reflected in lower adoption levels during the former season. The crude adoption rate appears to be positively associated with farm size. Tenancy appears to have a depressing effect on adoption and members of farmers' organisations are observed to have significantly higher adoption rates compared to non-members. It is observed that the propensity to adopt HYV is positively associated with farm size and membership in farmers' organisations; it is negatively associated with tenancy. It can be concluded that both the large- and medium-sized farms are more likely to adopt HYV compared to the small farms; owner operators are more likely to adopt compared to the tenants; and members of farmer organisations are more likely to adopt compared to the non-members. It also appears that innovativeness of a farmer is not related to his age and that on-farm labour availabili- is not a significant constrint to HYV adoption.

---

[1] This has particularly been observed during Boro season when 66% of the farms adopted HYV and grew them on more than one half of their rice area; during Aman when only 40% of the farms grow HYV on only 30% of their rice area, the inverse relationship between productivity and farm size appears to disappear.

Just literacy and lower primary education do not appear
to have any significant effect on either the farmer's propensity to adopt or on his intensity of adoption. However,
cultivators who have completed education up to primary level
and above are more likely to adopt HYV Aman compared to
illiterate farmers and the intensity of adoption is also
significantly higher for the former. It is interesting to
note that these differences have surfaced in the cultivation
of the riskier crop.

Intensity of HYV adoption is <u>negatively</u> associated with
farm size; it is <u>negatively</u> associated (during major crop season)
with tenancy; and it is <u>positively</u> associated with membership in
farmers' organisations.

It is observed that the positive relationship between size
and the crude adoption rate is neutralised by the negative relationship between size and the intensity of adoption with the
result that no definite pattern of relationship emerges between
size and index of participation. The higher crude adoption rate
among owner-cultivator is reinforced by their higher intensity
of adoption to produce a higher index of participation among
owner-cultivators compared to the tenants. Moreover, the owner
cultivators are observed to have a higher index of participation
compared to owner-cum-tenant farms. During Aman the owner-cum-
tenant category of farms (although for one season) also appear
to have a higher index of participation compared to that of the
tenant farms. However, during Boro the higher intensity of
adoption among tenant farms compared to owner-cum-tenant farms
results in higher index of participation for the tenants because
there is not much difference in the crude adoption rates of these
two farm categories during that season. Members of farmers'
organisations have both significantly higher adoption rates and
intensity of adoption and this has resulted in considerably higher
index of participation among member farms.

From the dynamics of HYV adoption it appears that the crude
adoption rates do not initially differ across farm size; however,
the adoption rates among large farmers tend to be higher in the long
run. Although this gap in adoption rate across farm size has
emerged over time, it is to be noted that the smaller farms have
also attained high adoption rates over time. Although the owner-
cultivators took the lead in HYV adoption, in the long run owner-cum-
tenant cultivators caught up with them. The tenant cultivators'
adoption rates were initially the lowest and they continued to be the
laggards over time. Insecurity of tenancy and the prevailing share-
cropping system continues to depress the tenant cultivators' rates of
HYV adoption over time. Members of farmers' organisations pioneer HYV
adoption and this lead over the non-member cultivators is maintained
over time. Finally, significant regional differences in HYV adoption
are observed. Under both a traditional agricultural setting and
under conditions of technological change, an _inverse_ relationship betwee
land productivity and farm size has been observed in conjunction with
constant returns to scale in Bangladesh agriculture. Rice output per
acre is _negatively_ associated with farm size and tenancy. With the
introduction of the green revolution technology, the _inverse_ relationships between productivity and farm size has weakened during Boro and
has disappeared during Aman. The inverse relationship between productivity and tenancy is still valid for Boro season (when HYV adoption
rate and intensity of adoption are high); during Aman this relationship
is no longer significant. Endogeneous man-made land improvement factor
such as irrigation and fertiliser application which affect land quality
have a significant _positive_ impact on land productivity. The above
inverse relationships have been established both in terms of pooled
(aggregate) and intravillage (disaggregate) data and as such the
conclusions are free of the statistical fallacy that one runs into by
relying on results obtained only from aggregate or pooled data. For
none of the crops the inverse relationship is fully universal.

After controlling for exogeneous land quality factors like soil
fertility and endogeneous land-improvement factors and the level of
mechanisation the above conclusions about the _inverse_ relationships
remain unaltered. Influence of the mode of labour use on land

productivity is not clear.   Significant regional differences in
productivity are observed for both crops.   Therefore, it can be
concluded that considerations of dynamic factors such as technological
change, need not reverse the policy implications that arise from the
static output gains to be achieved from a policy of land redistribution and other programmes favouring the small farm sector.   At the
very least it will not lead to a loss of output.   It can also be
concluded that share-cropping lessens the productive potential of
Bangladesh agriculture and output gains may be expected as a result
of a shift from share-cropping to farm ownership.   At the very least
such a policy measure will not lead to a decline in production.

Finally, a significant __negative__ relationship is observed between
cropping intensity and farm size both for adopters and nonadopters of
HYV.   No significant relationship is observed to exiwt between
__cropping intensity and tenancy__.   The __inverse__ relationship between
cropping intensity and farm size is most convincingly established
when it is seen that even after controlling for the availability and
use of irrigation water, this __inverse__ relationship remains unaltered
both under traditional agriculture and the new technology.   One
important reason for the existence of this negative relationship is
that the big farmers in the survey area were found to be engaged in a
multiplicity of channels of profit making and the cultivation of land
appears on their agenda only for a brief period.

Tenancy prevents the farms from being integrated into the monetised
market economy and tends to leave them in the subsistence sector.

Per acre labour intensity is __positively__ associated with land
improvement factors like irrigation, fertilisation and weeding; it
is __negatively__ associated with the use of mechanisation for preparatory
tillage.   The analysis of the operation-wise variations in labour
intensity is more revealing.   Per acre labour intensity is
positively associated with the dose of farmyard manure and nitrogen
fertiliser applied per acre.   Substitution of the other types of

chemical fertilisers for farmyard manure is labour-saving.
Labour input for weeding is positively associated with the
fertiliser/farmyard manure dose applied, with the frequency of
weeding and with the practice of row planting. Irrigation
method used is a significant determinant of the intensity of
labour-use in this operation. As expected, irrigation labour
input per acre is positively associated with the use of hand
tubewells, powered tubewells and indigenous methods whereas
the power pump (gravity flow) irrigation was not significant.
Use of mechanisation for preparatory tillage leads to the loss
of employment per acre and the average labour intensity appears
to be invariant to crop variety and agricultural season.

With the introduction of the green revolution technology
the inverse relationship between per acre labour intensity and
farm size is observed. Although this does not appear to be a fully
universal law, this feature occurs repeatedly in different
combinations of crop varieties, seasons and specific farm
operations in different parts of the country.

With the introduction of the green revolution technology,
the emergence of a significant depressing effect of land tenancy
on the intensity of labour use per acre is also observed. This
is established in terms of both pooled and intra-farm data. As
in the case of the inverse relationship between labour intensity
and farm size, this is not a fully universal law, although this
feature occurs in different combinations of crop variety, season
and specific farm operation in different parts of the country.
Dependence of labour intensity on the mode of labour use does not,
by and large, appear to be significant.

Over one half of the entire labour days applied on an acre of
rice crop cultivated is hired labour. Hired labour (number of
labour days as well as hired proportion of total labour) applied
per acre is positively associated with the percentage of the
cropped area under HYV. Per acre use of hired labour days does

not depend upon the level of mechanisation in preparatory tillage,
although the hired proportion of total labour use appears to be
positively related to it.   Hired labour (number of labour days a
as well as hired proportion of total labour) applied per acre
is _positively_ associated with farm size;   it is _negatively_
associated with land tenancy and with the number of adult
agricultural workers in the household.   Significant regional
differences in the per acre intensity of labour (hired and total)
use is observed.

The high degree of credibility accorded to both indigeneous
channels of communication ("neighbours") and government extension
agencies speaks of the need for greater utilisation of local
channels of communication through local leaders and village level
extension agencies.   Indigenous channels are particularly
important for smaller farms while the official extension agencies
appeared to be biased towards disseminating knowledge on HYV
technology to the larger farmers.   The mass media is virtually
ineffective.

Reliance on the noninstitutional sources of credit, in terms
of amount borrowed, and the number of borrowers, is very high and
this means the costs of borrowing from the rural capital market is
also high.   While the reliance and cost of working capital is
high for all cultivators, it is greater for the smaller farms.
The existence of these imperfections in the factor market is not
surprising.   Credit worthiness is positively related to farm size.
For example, gross income is positively related to farm size;
ownership of farm assets (which provides ability to offer collateral
for loans) is positively related to farm size;   and loan repayment
records particularly with respect to institutional sources is
better for larger farms, implying lower defaults on loans by this
group.

Between one-third and two-thirds of the fertiliser purchasers
obtain their supplies from the black market by paying between one

and a half to two and a half times the official dealer's price and
the fertiliser purchasers are reliant on the black market for the
bulk of their supplies.   While the cultivators as a whole are in
the grips of the black market, the smaller farms' dependency of this
costlier market is greater.

The proportion of rice output marketed is <u>positively</u> associated
with farm size;   it is <u>negatively</u> sssociated with tenancy.   Membership in farmers' organisations does not appear to have an. influence
on the marketable surplus.

The access to farmers' organisations appears to be quite unequal.
Larger farmers and medium sized farmers are more likely to participate
in farmers' organisations than smaller farmers;   owner-cum-tanants
are more likely to participate than pure tenants;   farmers with primary
or higher levels of education are more likely to participate than
others with none or lower-level education.

Policy Implications

From the above it is clear this it is often the small-farm sector
in Balgladesh that is most successful in achieving high levels of
production relative to modest inputs of capital and the scarcest
resource, land through the ample use of the country's abundant factor,
labour.   Moreover, it has been convincingly established that following
the introduction of the green revolution technology, it is precisely
the small agricultural holdings which can significantly increase labour-
use without sacrificing productivity.   The small-farm sector has also
demonstrated a capacity to achieve greater land utilisation through
the attainment of higher multiple cropping ratios both under traditional
agriculture and new technology.   Moreover, they take greater
advantage of the HYV technology as they are able to plant a higher
proportion of their acreage to HYVs.   This being feasible, once again,
through the liberal application of the family labour to such high labour
intensity crops.   The prevailing share-cropping system tends to depress
output and labour-use under both traditional farming and the new
technology;   it reduces labour-intensity particularly under the new
technology;   it also constrains the diffusion of innovations.   Therefor
the central policy implications of this study are that agrarian reform i
a vital policy instrument for achieving the combined objectives of

agricultural growth, employment generation and rural equity. It can
be concluded that sonsideration of dynamic factors, such as technological change, need not reverse the policy implications that arise from
the static output gains to be achieved from a policy of land redistribution (or programmes favouring the small farm sector)[1] and tenurial
reform.[2] On the contrary, from the important perspective of employment generation the above agrarian reforms assume special significance
with the introduction and wider dissemination of the green revolution
technology in Bangladesh. On the one hand, the proposed rural
development strategy calls for radical land redistribution measures
and tenurial reform (improvement in security of tenure, changes in
rental arrangements, transfer of ownership rights to the tiller, etc.)
on the other. In the event, such drastic agrarian reform measures
are not politically feasible, an alternative, though second best,
policy option would be to channel the improved farm inputs (HYV seeds,
fertilisers, pesticides, irrigation facilities, etc.) extension
services, knowledge on new technology and agricultural credit to the
existing small farm sector. In any event, urgent corrective action
is needed in this area as the evidence from this study shows that the
large-farm sector (either individually or as the co-operative "elite")
continues to gain preferential access to cheap credit and highly
subsidised modern farm inputs.

Diffusion of the HYV technology deserves priority consideration
not only because of the higher demand it generates for the use of
hired labour (important in view of the acute and growing level of
landlessness in the countryside). Mechanisation in land preparation
will have to be avoided as it leads to labour displacement without
providing any output advantages. Mechanisation in irrigation is
desirable as it permits multiple cropping and encourages the adoption
of HYVs through the creation of conditions favourable (availability of
controlled and assured water supply) to it. Labour-use in irrigation

---

[1] In a traditional agricultural setting land redistribution would promote
growth in productivity with equity and generate employment. Under
conditions of technological change land distribution will help attain the
objectives of generating more employment and promoting distributional
equity without sacrificing output.

[2] This measure will lead to simultaneous increases in output and employment; at the minimum it will either increase employment without reducing
output (green revolution situation) or it will raise output without loss of
employment (traditional agriculture).

could be further increased if emphasis is placed in the use of
irrigation methods like hand pumps, powered tubewells and indigeneous
techniques compared to low-lift pumps wherever such a choice is
feasible.

The high degree of credibility accorded to both indigeneous
channels of communication and government extension agencies speaks
of the need for greater utilisation of local leaders and village
level extension services.   Indigeneous channels are particularly
important for the small farm sector;   the government extension
agencies being biased towards disseminating knowledge on new
technology to the larger farmers.   Finally, mass media cannot be
relied upon as an effective channel of communication due to high
illiteracy and lack of access to radio sets.

Bibliography

1. Abdullah, Abu A., "Land Reform and Agrarian Change in Bangladesh", The Bangladesh Development Studies, Vol. IV No. 1, January 1976.

2. _____, "Formulating a Viable Land Policy for Bangladesh – What Do We Need to Know," The Bangladesh Development Studies, Vol. VI, No. 4, Autumn 1978.

3. _____, M. Hossain and R. Nations, "Agrarian Structure and the IRDP – Preliminary Considerations", The Bangladesh Development Studies, Vol. IV, No. 2, April 1976.

4. _____, "SIDA/ILO Report on Integrated Rural Development Programme, IRDP, Bangladesh", Bangladesh Institute of Development Studies, June 1974.

5. Ahmed, Iftikhar, "Appropriate rice production technology for Bangladesh", Agricultural Mechanisation in Asia (Tokyo), Vol. VIII, No. 4, Autumn 1977.

6. _____, "Technological Change, Agrarian Structure and Labour Absorption in Bangladesh Rice Cultivation," paper presented at the Seminar on Employment Expansion in South Asian Agriculture (Dacca), Government of Bangladesh – ILO/ARTEP (Bangkok), November 1979.

7. _____, "The Green Revolution in Bangladesh: Adoption, Diffusion and Distributional Questions", Geneva, ILO, 1975; mimeographed World Employment Programme Research Working Paper.

8. Asaduzzaman, M. and Faridul Islam, "Adoption of HYVs in Bangladesh: Some Preliminary Hypotheses and Tests", Bangladesh Institute of Development Studies, Research Report Series, New Series No. 23.

9. _____ and Mahabub Hossain, "Some Aspects of Agricultural Credit in Two Irrigated Areas in Bangladesh", Bangladesh Institute of Development Studies, Research Report Series, New Series No. 18, October 1974.

10. Bardhan, Pranab K., "On labour absorption in South Asian rice agriculture with particular reference to India", Labour Absorption in Indian Agriculture: Some Exploratory Investigations, Asian Regional Team for Employment Promotion, ILO, Bangkok, November 1978.

11. _____, "Size, Productivity, and Returns to Scale: An analysis of Farm-Level Data in Indian Agriculture", Journal of Political Economy, Vol. 81, No. 6, November/December 1973.

12. _____ and T.N. Srinivasan, "Cropsharing Tenancy in Agriculture: A Theoretical and Empirical Analysis", American Economic Review, Vol. LXI, No. 1, March 1971.

13. Berry, R. Albert and William R. Cline, "Agrarian Structure and Productivity in Developing Countries", Johns Hopkins University Press, 1979.

14. Bhati, U.N., "Some Social and Economic Aspects of the Introduction of New Varieties of Paddy in Malaysia – A Village Case Study", United Nations Research Institute of Social Development, Report No. 76.8, Geneva.

15. Binswanger, Hans P. "The Economics of Tractors in South Asia: An Analytical Review", Agricultural Development Council (New York) and International Crops Research Institute for the Semi-Arid Tropics (Hyderabad, India), 1978.

16. Blaug, M., "Education and the Employment Problem in the Developing Countries", ILO, Geneva, 1973.

17. Bose, A.B. and M.B. Jain, "Relative Importance of Some Socio-Economic Factors in the Adoption of Innovations", Indian Journal of Social Research, Vol. 7, No. 1, April 1966.

18. _____ and P.C. Saxena, "The Diffusion of Innovations in a Village in Western Rajasthan", Eastern Anthropologist, Vol. 18, No. 3, 1965.

19. Chandra, Nirmal K., "Farm efficiency under semi-feudalism: a critique of marginalist theories and some Marxists formulations", Economic and Political Weekly, Vol. IX, Nos. 32-34 (Special Number) 1974.

20. Chattopadhyay, M. and Ashok Rudra, "Size-productivity revisited", Economic and Political Weekly, Vol. XI, No. 39, September 25, 1976.

21. Chaudhri, D.P., "Education, Innovations and Agricultural Development", (Croom Helm, London, 1979.

22. Cheung, S. "Private Property Rights and Share-cropping", Journal of Political Economy, Vol. 76, No. 6, November/December, 1968.

23. Chinappa, B.N., "Adoption of the new Technology in North Arcot District" in B.H. Farmer (ed.), "Green Revolution? Technology and Changes in Rice-Growing Areas of Tamil Nadu and Sri Lanka", London, Macmillan, 1977.

24. Cramer, J.S., "Empirical Econometrics", North Holland 1969.

25. Cummungs, J.T., "The Supply Responsiveness of Bangaleo Rice and Cash Crop Cultivators", Bangladesh Development Studies, Vol. II, No. 4, October 1974.

26. Dasgupta, B., "Agrarian Change and the New Technology in India", United Nations Research Institute for Social Development, Geneva, 1977.

27. Dias, Hiran D. "A Land Reform Policy (for Sri Lanka) in the Context of HYVs", IDS (Sussex) - ARTI (Colombo) Seminar on the Economic and Social Consequences of the Improved Seeds, Kandy, Sri Lanka, April - May 1973.

28. Faidlay, L. and M. Esmay, "Introduction and Use of Improved Varieties: Who Benefits" in Robert D. Stevens, et al "Rural Development in Bangladesh and Pakistan," The University Press of Hawaii, 1975.

29. Finney, D.J., "Probit Analysis", Cambridge University Press, 1971.

30. Ghai, Dharam, et al (editors), "Agrarian Systems and Rural Development", Macmillan, 1979.

31. Government of Bangladesh, "Guidelines of the Second Five-Year Plan (1980-85)", Planning Commission, General Economics and Evaluation Division, Dacca, June 1979.

32. _____, "Statistical Pocket Book of Bangladesh 1978", Ministry of Planning, Statistics Division, Bangladesh Bureau of Statistics.

33. _____, "The First Five Year Plan 1973-78", Planning Commission, Dacca, November 1973.

34. _____, "The Revised Two Year Plan (1978-80) of the Agricultural Sector, Ministry of Agriculture, Dacca, April 1978.

35. Government of India, "Report on Evaluation of the High Yielding Varieties Programme, Kharif 1968", Planning Commission, Programme Evaluation Organisation, Publication No. 67, June, 1969.

36. Grabowski, Richard, "The Implications of an Induced Innovation Model", Economic Development and Cultural Change, Vol. 27, No. 4, July 1979.

37. Griffin, Keith, "The Political Economy of Agrarian Change", Macmillan 1974.

38. Hossain, Mahabub, "Agrarian Reform and Rural Development: A Review of Recent Experiences in Selected Asian Countries", Bangladesh Institute of Development Studies, (mimeo) July 1979.

39. _____, "Farm Size, Tenancy and Land Productivity: An Analysis of Farm Level Data in Bangladesh Agriculture", The Bangladesh Development Studies, Vol. V, No. 3, July 1977.

40. Ishikawa, S., "Labour Abosrption in Asian Agriculture" Asian Regional Team for Employment Promotion, ILO, Bangkok, June 1978.

41. Jabbar, M.A., "Relative Production Efficiency of Different Tenure Classes in Selected Areas of Bangladesh", The Bangladesh Development Studies, Vol. V, No. 1, July 1977.

42. Khan, Azizur Rahman, "Poverty and Inequality in Rural Bangladesh", Poverty and Landlessness in Rural Asia, International Labour Office, Geneva, 1977.

43. _____, "The Comilla Model and the Integrated Rural Development Programme of Bangladesh: An Experiment in Cooperative Capitalism", World Development, Vol. 7, No. 4/5, April/May 1979.

44. _____ and Iftikhar Ahmed, "Labour input in rice cultivation: a preliminary analysis of some Bangladesh data", paper presented at the Technical Workshop on Labour Absorption in Asian Agriculture, ILO's Asian Regional Team for Employment Promotion, Bangkok, August 1978, (mimeo).

45. Khan, Mahmood H. and Dennis R. Maki, "Effects of Farm Size on Economic Efficiency: The Case of Pakistan", American Journal of Agricultural Economics, Vol. 61, No. 1, February 1979.

46. Kmenta, Jan, "Elements of Econometrics", Macmillan 1971.

47. Lau, L.J. and P.A. Yotopoulos, "A Test for Relative Efficiency and Application to Indian Agriculture", The American Economic Review, Vol. LXI, No. 1, March 1971.

48. Librero, Aida, "Socio-Economic Implications of High Yielding Varieties in the Philippines", United Nations Research Institute for Social Development, Geneva, Report No. 76/C,1,1974.

49. Lockwood, Brian, P.K. Mukherjee and R.T. Shand, "The High Yielding Varieties Programme in India", Part 1, The Australian National University and the Planning Commission of India, Delhi, 1971.

50. Lowdermilk, Max K., "Preliminary Report of the Diffusion and Adoption of Dwarf Wheat Varieties in Khanewal Tehsil (Pakistan), Cornell University, Ithaca 1971 (mimeo).

51. Mellor, John W., "The New Economics of Growth: A Strategy for India and the Developing World," Cornell University Press, 1976.

52. Newberry, D., "Cropsharing Tenancy in Agriculture: Comment", The American Economic Review, Vol. LXIV, No. 6, December 1974.

53. Palmer, I., "The New Rice in Asia: Conclusions from Four Country Studies", United Nations Research Institute for Social Development, Geneva, 1976.

54. _____, "The New Rice in Indonesia", United Nations Research Institute for Social Development, Geneva, 1977.

55. Peek, Peter and Pedro Antolinez, "Migration and the Urban Labour Market: The Case of San Salvador", World Development, Vol. 5, No. 4, 1977.

56. Reddy, S.K. and J.E. Kivilin, "Adoption of HYV in Three Indian Villages", Research Report No. 19, National Institute of Community Development, Hyderabad, India, May, 1968.

57. Rochin, Refugio I., "A Study of Bangladesh Farmers' Experiences with IR-20 Rice Variety and Complementary Production Inputs," The Bangladesh Economic Review, Vol. 1, No. 1, January 1973.

58. Sau, Ranjit, "Land Utilisation: A Note", Economic and Political Weekly, Vol. XI, No. 36, September 4, 1976.

59. Schluter, M. and John W. Mellor, "New Seed Varieties and the Small Farm", Economic and Political Weekly, Vol. VII, No. 13, March 25, 1972.

60. Sidhu, Surjit S., "Relative Efficiency in Wheat Production In the Indian Punjab", The American Economic Review, Vol. LXIV, No. 4, September 1974.

61. Standing, Guy, "Female Labour Supply in an Urbanising Economy", in Guy Standing and Glen Sheehan, "Labour Force Participation in Low Income Countries", International Labour Office, Geneva, 1978.

62. The World Bank, "Bangladesh: Current Trends and Development Issues", South Asia Regional Office, Washington D.C., March 1979.

63. Theil, H., "Principles of Econometrics", J. Wiley and Sons, New York, 1971.

Appendix

In the analysis of a cultivator's propensity to adopt HYV (Tables 4 and 5), the dependent variable we used is a zero-one binary variable. It takes the value of one if the individual has adopted HYV and a value of zero if he has not. The parameters of a regression model with such a dependent variable cannot be estimated by the ordinary least squares method, because the disturbance term in the equation will not be distributed normally; it will assume a discrete distribution. We can illustrate this with an example.[1] Consider the linear equation:

$$(1) \quad Y_i = a + b\, x_i + e_i$$

where e is the disturbance term and x_i is an explanatory variable, e.g. farm size of the respondent. The dependent variable Y_i takes the value of either 1 or 0 (1 if the individual cultivator has adopted HYV, 0 if otherwise). This implies that for a given x_i, the disturbance term, e_i, also takes only two values: $-(a + b\, x_i)$ and $1 - (a + b\, x_i)$. Thus e_i is not normally distributed but has a discrete distribution. It can be shown[2] that e_i is heteroskedastic since its variance depends on the mathematical expectation of the dependent variable. Because of the heteroskedastic nature of the disturbance term, the least squares estimators of a and b will not be efficient.

One way to deal with such a problem is to make a transformation on the dependent variable; and then apply the maximum likelihood method for the estimation of parameters. Probit and logit analysis are concerned with such a transformation of the dependent variable.[3]

---

[1] For a discussion of dummy dependent variables see [63 and 46].

[2] See [46].

[3] Probit is derived from the words 'probability units'. While logit is taken from the words 'logistic units' (units from a logistic function). Logit owes its name to the relationship with the logistic function.

Applying the probit[1] monotonic transformation to the
probability $p(x_i)$ [where $p(x_i)$ is the probability that
an individual cultivator adopted HYV given x_i, the
size of his farm], its transform increases from
$-\infty$ to $+\infty$ when $p(x_i)$ increases from zero to one; the
probit specification is:

$$(2) \quad p(x_i) = f(a + bx_i) = \frac{1}{\sqrt{2\pi}} \int_{-\infty}^{a + bx_i} \exp\left[-\frac{1}{2} u^2\right] du$$

which is equivalent to: $\quad f^{-1}(p(x)) = a + bx_i$,
f being the cumulated distribution function of the standardized normal variate: $a + bx_i$. From the above, it is
assumed that the farm size of an individual cultivator who
has adopted HYV varies with each individual. It is further
assumed that x_i is distributed normally.

In the above paragraph we observe that in probit
analysis f is assumed to be normally distributed. This
assumption is rather strong for the problem we are investigating. For this reason we use another transformation, the
logit. Logit analysis involves a more direct transformation.
The dependent variable is defined as $\frac{p}{1-p}$, where p is the
probability of HYV adoption. Then specified in log linear
form:

$$(3) \quad \log\left(\frac{p}{1-p}\right) = a + b \log x$$

and solving for p

$$(4) \quad p = 1/[1 + e^{-a} x^{-b}].$$

The left side of (3) which is known as the logit of HYV
adoption is a transformation of the probability which is
equal to $-\infty$ at p=0, 0 at p=.5, ∞ at p=1. In other words, the
logit transformation provides for p moving from $-\infty$ to $+\infty$ as y,
the transformed dependent variable moves from 0 to 1. It is
also clear that the approach can be extended when more explanatory variables are needed.

---

[1]Actually, probit analysis includes a number of methods each
with a different transformation [29]. Originally these methods
were introduced in biology to measure the response to the stimulus
of a drug or other chemical preparation. It has subsequently been
applied in the social sciences - for example, in the analysis of
the propensity to participate in the labour force [60] and in the
analysis of the propensity to be employed in the urban formal or informal sector in relation to the numerous socioeconomic character-
istics of the migrant worker [55].